SENSITIVE WITNESSES

SENSITIVE WITNESSES

Feminist Materialism in the British Enlightenment

KRISTIN M. GIRTEN

STANFORD UNIVERSITY PRESS
Stanford, California

Stanford University Press
Stanford, California

This book has been partially underwritten by the Peter Stansky Publication Fund in British Studies. For more information on the fund, please see www.sup.org/stanskyfund.

Printed and bound by CPI Group (UK) Ltd, Croydon, CR0 4YY

Library of Congress Cataloging-in-Publication Data
Names: Girten, Kristin M., author.
Title: Sensitive witnesses : feminist materialism in the British Enlightenment / Kristin M. Girten.
Description: Stanford, California : Stanford University Press, 2024. | Includes bibliographical references and index.
Identifiers: LCCN 2023046088 (print) | LCCN 2023046089 (ebook) | ISBN 9781503633032 (cloth) | ISBN 9781503637696 (ebook)
Subjects: LCSH: English literature—Women authors—History and criticism. | Philosophy of nature in literature. | Materialism in literature. | Philosophy and science—Great Britain—History. | Authors, English—Philosophy—History. | Enlightenment—Great Britain.
Classification: LCC PR113 .G57 2024 (print) | LCC PR113 (ebook) | DDC 820.9/9287—dc23/eng/20231114
LC record available at https://lccn.loc.gov/2023046088
LC ebook record available at https://lccn.loc.gov/2023046089

Cover design: Lindy Kasler
Cover painting: Philip Reinagle, *The Blue Passion Flower, for Robert John Thornton's Temple of Flora (London, 1799–1812)*, 1800; Yale Center for British Art, Paul Mellon Collection

To Nick

CONTENTS

ACKNOWLEDGMENTS

As you will discover in the pages of this book, Lucy Hutchinson managed to produce her translation of Lucretius's *De Rerum Natura* while her children were taught by their tutors in the same room. I understand well the challenges and opportunities that come from balancing a life as a scholar with a life as a devoted mother. Such has been my experience while researching and writing *Sensitive Witnesses*, my first monograph. My twin children, Miles and Lyndon, have accompanied me each step along the way, even when they weren't aware of doing so. I have learned so much about life and love, as well as what it means to be a sensitive witness, from them. As they have matured, they have become enthusiastic supporters of my scholarship. I am deeply grateful for their love and encouragement.

As a result of my devotion to my children and to my students, this book has been in development for many years. I believe it is all the better for it. I am grateful for my home institution, the University of Nebraska at Omaha (UNO), which has allowed me to develop as a scholar, and to pursue my research, at my own pace. Thanks to its support, I was successful at earning tenure and establishing myself in my field of eighteenth-century and Romantic studies with articles alone, which I published in three of my favorite journals: *Eighteenth-Century Studies*, *Eighteenth Century: Theory and Interpretation*, and *Eighteenth-Century Fiction*. I am grateful to the editors of these journals whose valuable support and guidance helped me to refine my scholarship. I want to extend special thanks to Rivka Swenson and Manushag N. Powell who co-edited the special issue of *ECTI* on "Sensational Subjects" in which my article "Charlotte Smith's Tactile Poetics" appeared. Their feedback was invaluable and helped me maximize the clarity, meaning, and impact of my essay. In addition, I am also profoundly grateful to my UNO colleagues, Sue Maher and John Price, who have been mentors and friends to me when I have needed them most. Sue has been an inspiration to me in her career as both a scholar and an administrator. John has been a close friend who shares my appreciation for the value of "environmental humanities" and commitment to giving ineffable experiences the attention they are due. I would also like to express my appreciation for Carey Wolfe and the late Tim Wiles who were great sources of inspiration to me in my earliest days as a scholar. I am not sure I could be where I am today had I not received Tim's encouragement. He will always be missed.

Charlotte Smith's poem *Beachy Head* has been a driving force in my work since the late 1990s when I was completing my MA in English at the University of Colorado at Boulder. I want to give special acknowledgement to my mentors there, Jeffrey Robinson and Timothy Morton, who were the first to help me see how generative and fulfilling literary study could be. Jeffrey enlivened me to the beauty of describing poetry in its exquisite detail and to the power of women Romantic poets who succeeded at making flights fancy as transporting and meaningful as works of the imagination. Tim has been both a mentor and friend. I would also like to extend my gratitude to my Rutgers University community of English faculty and graduate student colleagues: Michael McKeon, Paula McDowell, Colin Jager, Jonathan Kramnick, Carolyn Williams, Richard Dienst, Drucilla Cornell, Kathy Lubey, James Mulholland, Danielle Bobker, Richard Squibbs, Kathryn Steele, Sarah Ellenzweig, Deborah Allen, Melanie Holm, Madhvi Zutshi, Ray Ricketts, Joseph Chaves, John Shanahan, Mark Vareschi, Tina Lupton, Barrett Kalter, Laura Diehl, and Brian Norman. Thanks also to Greg Nichol, Jen Rector, and Dave Pantos for their friendship and support over the years. In addition, I would like to express my gratitude to those faculty and students at Princeton in the early aughts who have significantly inspired my scholarship: Jonathan Lamb, Dahlia Porter, Ingrid Horrocks, Gabriel Cervantes, Hollis Robbins, and Denise Gigante.

I have been fortunate to share my work with colleagues at numerous conferences, including the American Society of Eighteenth-Century Studies, the North American Society for the Study of Romanticism, and the Association for the Study of Literature and Environment. I am particularly appreciative of the support I have received in the development of this book from Helen Thompson, Wendy Lee, Tita Chico, Susan Carlile, Sarah Kareem, Courtney Weiss Smith, Danielle Spratt, David Alff, and Rick Barney. Thanks also to the institutions that have enabled my research for this book: specifically, the British Library, the Huntington Library, the Indiana University Lilly Library, and the UNO Criss Library. A special thank you to Katie Bishop, my area librarian at my home institution, UNO. I am grateful for the grant support I have received to support my work from the Lilly Library and the UNO University Committee for Research and Creative Activity (UCRCA). In addition, I have received support through an NEH Summer Seminar, "The Aesthetics of British Romanticism, Then and Today," facilitated by Stephen Behrendt at the University of Nebraska at Lincoln. Steve was an important inspiration and supporter when I was first researching "the material sublime" as both an Enlightenment and Romantic phenomenon. Through this NEH Summer Seminar, I became acquainted with several Romantic scholars who have been valuable sources of inspiration to me. In particular, I would like to extend my gratitude to Allison Dushane, Soledad Caballero, Jason Goldsmith, and Seth Reno, all of whom remain friends and scholarly confidants. Special thanks to Seth, who recently included my article on sensitive witnessing and Jeff VanderMeer's *Borne* (2017) in his collection *The Anthropocene: Approaches and Contexts for Literature and the Humanities* (Routledge, 2022).

I am deeply grateful to Stanford University Press and its editors. In particular, I would like to extend my thanks to Erica Wetter and Caroline McKusick. Erica advocated for my book to the press and was instrumental in helping me secure an advance contract. I have relied on Caroline throughout the review process and while preparing the manuscript for print. I feel extremely fortunate to have been able to count on them for support. They both embody that rare combination of kindness and careful attention to detail that one wants in an editor.

I dedicate this book to my husband, Nick Monk, who has believed in me since we first became friends many years ago. Though he and I research and write in different literary fields and about different continents, he is my most valued interlocutor. He gets me and my work in ways no one else can and, consequently, is able to help me express its meaning and significance even when they aren't yet clear even to me. He shares my love of all things sublime (both of the vast and of the small) as well as my affection for creatures and "the natural world." He is the best close reader I know and also my favorite person to go birding and snorkeling with. I will always cherish my memories of our travels together and particularly that time at the foot of Beachy Head when he asked me to marry him. Our life together is abundant with "spots of time." Nick, you are and will always be the love of my life.

SENSITIVE WITNESSES

INTRODUCTION

IN 1759, JONATHAN SWIFT sparked considerable outrage and debate with the publication of *A Modest Proposal*, a work now regularly presented to students as a quintessential example of early modern satire. Though the precise target of Swift's *Proposal* is notoriously difficult to pin down, as is the case with the majority of his works, its title gives us an important clue.[1] Its satire is directed at "modesty," embodied by the *Proposal*'s speaker. To discern what this parody entails, we must ask: what exactly does Swift mean by the term *modest*? At the time of the work's composition and publication, this term was variable in usage, shifting according to the gender of the subject to whom it was applied. For women, modesty suggested one who was "decorous in manner and conduct; not forward, impudent, or lewd; demure; . . . scrupulously avoiding impropriety or vulgarity in speech or behaviour."[2] For men, in contrast, it implied "[h]aving a moderate or humble estimate of one's own abilities or achievements; disinclined to bring oneself into notice; becomingly diffident and unassuming; not bold or forward."[3] As multispecies feminist theorist Donna Haraway observes, "[f]emale modesty was of the body; the new masculine virtue [of modesty] had to be of the mind."[4] Given that the speaker of Swift's *Proposal* is male, we may assume that it is the modest man whom the work parodies and thus satirizes.

Haraway characterizes masculine modesty as a "new virtue" because of how central it would become to the new empirical philosophy—what we would now call science—of the Enlightenment era. With his 1620 *Novum Organum*, the forefather of modern science, Francis Bacon, developed a distinctive philosophic orientation to nonhuman nature that perceived man as both separate from and dominant over the material world. It is this perception that informs the masculine modesty of the Enlightenment era; therefore, the Enlightenment philosopher has come to be known as a "modest witness." Whereas women were expected to preserve the modesty of their bodies—by embodying virtue and thereby preserving their chastity—men were expected to preserve the modesty of their minds. They were to do so by remaining detached from and transcendent over the material world. Consequently, while the modest woman was to be courteous, polite, and demure in her actions, the modest man was to be aloof, sober, and neutral in his thinking. It is precisely such masculine modesty that the male speaker of Swift's *Modest Proposal* performs and that the work thus satirizes. In *A Modest Proposal*, Swift shows that modesty—specifically, the modesty entailed by "calculating the financial worth of a population" or what was then known as "political arithmetic"—was not only unreliable as a means by which to ensure the authority of one's observations and analysis; it was also a source of arrogance and inhumanity as well as utter cluelessness.[5] As such, it could produce the most ridiculous of ideas. In the case of Swift's *Modest Proposal*, it inspires the simultaneously ridiculous and abhorrent promotion of cannibalism—specifically, the eating of children: "A young healthy child well nursed, is, at a year old, a most delicious nourishing and wholesome food, whether stewed, roasted, baked, or boiled; and I make no doubt that it will equally serve in a fricassee, or a ragout."[6]

As Steven Shapin and Simon Schaffer explain in their foundational work *Leviathan and the Air-Pump*, "a man whose narratives could be credited as mirrors of reality was a *modest man*; his reports ought to make that modesty visible." He did so by using the form of "the experimental essay," by adopting a "naked way of writing," and by "speak[ing] confidently of matters of fact."[7] These techniques were perceived to allow the scientist to establish and maintain his Baconian modesty. In 1663, they were institutionalized in the Royal Society's statutes thus: "In all Reports of Experiments to be brought into the Society, the matter of fact shall be barely stated, without any prefaces, apologies, or rhetorical flourishes; and entered so in the Register-book, by order of the Society."[8] "In other words, as Tita Chico observes, they were to use "language that is emphatically not literary."[9] Over the next century or so, such modesty would come to be known as the scientist's *objectivity*, his "view from nowhere."[10] The scientist's basic techniques for achieving modesty, and thereby becoming an "independent observer," have persisted into the present day.[11]

The philosophical method that Bacon pioneered was distinguished from the Aristotelian scientific tradition that preceded it by its reliance on experience, experiment, direct observation, and induction. Whereas previous scientific practices regularly tended to be primarily theoretical, speculative, or discursive, Bacon's was fundamentally *empirical* in nature. With his innovation came new challenges. Bacon was particularly concerned with what he called "the idols of the mind." Bacon claims that, "beset" by the idols, "the human understanding is like a false mirror, which, receiving rays irregularly, distorts and discolours the nature of things by mingling its own nature with it." The solution he provides is for the scientist to "free . . . and cleanse" their "understanding."[12] Bacon suggests that modern science requires that its practitioners purify themselves so they may be set apart and, thus, unencumbered by the very material world they seek to explore and understand. It is this indifferent neutrality, or modesty, that would eventually evolve into what we now refer to as "objectivity." Still regularly cited as the forefather of "the scientific method," Bacon has had an enduring legacy. Objectivity persists as a requirement of reliable scientific investigation. The scientist, then and now, is regularly expected to be an "independent observer" who resists the temptation to intermingle with, and thus become swayed by, their objects of study.

Women were, in the Enlightenment era, regularly understood to be highly sensitive and thus overly susceptible to external stimuli as well as fundamentally and thus unavoidably associated with materiality. Consequently, they were perceived as unequipped to achieve the kind of masculine modesty that their male philosophical counterparts strove to achieve. To cite a representative example, in 1708, the *British Apollo* replies thus to a reader who had asked "whether women were as capable of learning as men": women "are cast in too soft a mould, are made of too fine, too *delicate* composure to endure the severity of study, the drudgery of contemplation, the fatigue of profound speculation."[13] Throughout the Enlightenment, "sensitive," "tender," and "womanish" were close synonyms for "delicate," and these terms were regularly used in a pejorative sense.[14] According to physician Robert Whytt, "A DELICATE or easily irritable nervous system, must expose a person to various ailments, from causes affecting either the body or mind, too slight to make any remarkable impression upon those of firmer and less sensible nerves."[15] As this passage illustrates, regularly perceived as delicate creatures, women were thought to be particularly open and vulnerable to their environment ("causes affecting either the body or mind").[16] It is no surprise then that Baconian modesty, distinguished by a practice of separating oneself from one's environment, was generally viewed as unavailable to women. The distance required by masculine modesty was a privilege generally deemed incompatible with being a woman.

Women's identification with materiality was even seen to make them, along with the enslaved and the poor, *less human* than their male counterparts. As scholars regularly acknowledge, Locke's theory of the "property in one's person" provides an important foundation for the emergence and development of the modern subject. This theory, premised as it is on alienation, contributes significantly to the "buffered" nature of the modern self while also closely identifying Lockean subjectivity with possessive individualism.[17] The key passage comes from Locke's *Second Treatise,* published anonymously in 1689. It begins: "Though the earth, and all inferior creatures, be common to all men, yet every man has a *property* in his own *person:* this no body has any right to but himself."[18] Here, while depicting nonhuman nature as a commons, Locke nevertheless proceeds to separate individual men—and their "persons"—out from the commons. In so doing, he alienates men from one another. A man's ownership of his person fundamentally estranges him from his fellow man. Furthermore, he alienates men from themselves: a man may only possess "his own person" if he and "his person" are distinguished and thus estranged from one another. As many scholars have noted, this alienation was instrumental in the birth and development of the modern subject.[19] It ensures that, to return to Locke, "[t]he *labour* of his body, and the *work* of his hands . . . are properly his"—his to use to "exclude . . . the common right of other men" in the state of nature.[20] By so excluding "the common right" in what amounts to an act of "inclosure," man makes that which originates in the commons useful and thus valuable. Perhaps even more importantly, though, he asserts his own personal sovereignty over his environment as well as himself.[21] The property in one's own person thus ensures man's integrity as a subject as well as his freedom and power.

How Locke's theory of property, and the version of modern subjectivity it produces, applies to women has been a source of controversy from the date of its publication. As Ruth Perry observes, if "the definition of an individual" is "a property-owning being independent of all other property-owning beings—it is a definition which, in seventeenth-century England, deliberately excluded adult women." According to Perry, "[t]he requirement that a citizen own property in his own person is the crucial move in Locke . . . by which women were excluded from their place in the polity. . . . Their persons have invariably been understood to be the property of their fathers, husbands, or masters; their labor has been understood to be at the disposal of their families."[22] In the words of early modern Tory feminist Mary Astell, "[H]ow much soever Arbitrary Power may be dislik'd on a Throne, not *Milton* himself wou'd cry up Liberty to poor *Female Slaves*, or plead for the Lawfulness of Resisting a Private Tyranny."[23] Rape narratives accentuate the serious challenges women faced in claiming and maintaining their property in their person. As Leslie Richardson demonstrates, in the context of domestic fiction, "[t]he sexual act . . . functions as a symbolic occupation, similar to digging a clod of earth or cutting down a tree in staking a claim to real property." To document stolen sex is thus to "emphasiz[e] the tenuousness"—if not the impossibility—of "a woman's self-ownership."[24] Unable to own themselves, women could not even claim to be fully human. It is therefore no wonder that even the most privileged women struggled to assert themselves as philosophers.

In *Sensitive Witnesses*, however, I show how a group of female author-philosophers in the Enlightenment era managed to assert their philosophical authority by harnessing, rather than denying as the modest witness does, their association with materiality.[25] I argue that these author-philosophers counter the promotion of philosophical masculine modesty with an assertion that closeness to, rather than distance from, matter makes for good empirical observation and, thus, successful knowledge-making. Moreover, I assert that these author-philosophers variously drew on and/or contributed to contemporary materialisms that eschewed man's separation from matter.[26] They encourage a touching nature, according to which humanity is experienced as continuous with instead of alienated from its environment. Thus, they preview what has come to be known, in the twenty-first century, as "the new materialism," which is defined by a recognition of humanity's "trans-corporeal" "entanglement" with matter.[27] Transforming the liability of their gender into an asset, they envision a "material feminism" whereby their association with matter becomes an asset rather than a curse.[28] Touching nature, these women author-philosophers come to know nature in ways that one separate from and transcendent over nature never could.

Sensitive Witnesses calls for a revision to our understanding of the Enlightenment era and the philosophical methods for which it is known. I join scholars such as Helen Thompson, Tita Chico, Al Coppola, Alexander Wragge-Morley, and Peter Hanns Reill who complicate the "modesty" associated with the Age of Enlightenment.[29] Though early modern empirical philosophers claimed to reject "the paint, of art," nevertheless, as Chico demonstrates, "early science formulated itself through literary knowledge" and thus had recourse to imagination.[30] In other words, they were not as *modest* as their polemics and publicity purported them to be. In fact, as Wragge-Morley explains, "scientists of the early modern period and beyond recognized the epistemic potential of seemingly subjective states, from the passions of desire to the kinds of instant recognition arising from perceptual habit."[31] In *Sensitive Witnesses*, I show how a group of pioneering female authors also found philosophical value in "subjective states."

These philosophers who contributed to an alternative history of sublime knowledge-making enabled by the practice of sensitive witnessing prioritized knowledge gained through observation, experience, and experiment so profoundly embodied and engaged that it defied separation between scientist and specimen, subject and object. A central principle of Bacon's "Great Instauration" was a focus on the value of experience. In his "Dedicatory Epistle" to *Novum Organum*, Bacon writes, "[A]t last, after so many ages of the world, philosophy and the sciences may no longer float in the air, but rest upon the solid foundations of every kind of experience properly considered." He then goes on to say, "I have supplied the Instrument; but the material must be sought in things themselves."[32] With the new philosophy, experience—particularly sensory encounters with "things themselves"—was brought front and center as the best means of achieving knowledge and pursuing discovery. However, though Bacon emphasized the value of such sensory encounters with the material world, he simultaneously encouraged his followers to purify themselves of matter and thus to distinguish themselves from their material environments in preparation for their empirical projects. In contrast, the proponents of sensitive witnessing that *Sensitive Witnesses* explores challenged the legitimacy and feasibility of such purification. They instead insisted on the unavoidability of humanity's entanglement with matter and even derived their philosophical authority from it. *Sensitive Witnesses* explores how these female philosophers substantiated such entanglement and the practice of sensitive witnessing it informed. I argue that, as sensitive witnesses, these pioneering author-philosophers develop a distinctive approach to knowledge-making that is distinguished by sublimity.

Though Bacon promoted modesty and thus a clear separation between observer and observed, recent scholars have demonstrated that, nevertheless, early modern philosophical practices regularly capitalized on scientists' own material entanglement and, in so doing, violated and thus belied such separation. Contrary to the impression that much of the rhetoric around how science should be pursued and presented gave, intermingling with matter was a key dimension of early modern philosophers' empirical practice. *Sensitive Witnesses* demonstrates how various British female philosophers of the Enlightenment era used this often underplayed dimension of the new science to claim access to philosophical authority. In her book *Fictional Matter*, Helen Thompson challenges the modesty that Shapin and Schaffer and others have associated with the new science as she argues that Enlightenment era "empiricism institutes not separation but relation."[33] Exploring the empirical methods and principles of Robert Boyle (1627–1691), whom she presents as an "avatar of modern science," Thompson shows that corpuscular philosophies of matter readily persuaded new scientists such as Boyle to "define . . . empirical experience as the porosity of texture that admits . . . subtle . . . entering things."[34] Such a definition of experience runs directly counter to the gentlemanly modesty that Shapin and Schaffer associate with early modern empirical practice, as porosity fundamentally contradicts and thus precludes the buffering and detachment of the scientist that is supposedly distinctive of such modesty.

According to Thompson, "[t]he scientific intervention that galvanizes empiricism is not the split of sensed qualities from objects" as theories of modest witnessing would have us perceive. Rather, it is "the provenance of ideas as 'impressions' stimulated by environing bodies."[35] Thus, contrary to those who view Enlightenment era "objective" science as having set the foundation for man's enduring sense of alienation from and consequent destruction of the environment, Thompson's study shows that "[t]he shibboleth of 'scientific objectivity' . . . cannot accommodate the role of corpuscular power in eighteenth-century chemistry and empiricism."[36] Indeed, as Thompson demonstrates, Enlightenment era philosophical practice profoundly revealed humanity's impressionistic receptiveness to the environment and thereby made any assertion of one's separation from matter if not impossible then necessarily complicated: "Chemists like Boyle and Newton, as well as dyers, painters, gem and glassmakers, miners, metallurgists, brewers, tanners, submarine and shipbuilders, and other artificers, really *made* things with corpuscles (they also suffered from ingesting, breathing, and touching them)."[37]

It was not just in the laboratories of chemists like Boyle and Newton that corpuscular philosophy emphasized man's material openness to, and thus entanglement with, the environment. Works by moral philosophers such as

John Locke and David Hume did not only attest to such openness and entanglement but regularly theorized and thus helped emphasize it. Consequently, there are valuable connections to be made between such works and those by the female author-philosophers *Sensitive Witnesses* explores. Various scholars have recognized the influence of Boyle's corpuscularian chemistry on John Locke, and their scholarship invites an appreciation of how Locke's moral philosophy complicates, if not forecloses upon, masculine philosophical modesty.[38] Thompson's analysis of Boyle's influence on Locke is particularly relevant to the present study given her focus on the literary consequences of this influence. As Thompson explains, Locke portrays as "delusory" "the figural boundaries of things as humans perceive them"—including humans themselves. Consequently, for Locke as for Boyle, the detached modesty and buffering of the self that Bacon and his followers (including Boyle himself) recommend can only ever be aspirational and thus a performed fiction rather than a material fact.[39] In truth, according to Thompson's persuasive analysis, the entanglement between observer and observed becomes even more profound in Locke's *Essay Concerning Human Understanding* than it is in Boyle's corpuscularian chemistry as Locke denies "Boyle's distinction between mechanical affections or texture, which compose a thing's insensible but real physicality, and sensible qualities, which texture effects as ideas 'in us.'"[40] Whereas "Boyle divides mechanical affections in things from sensible qualities in persons," Locke in contrast "differentiates primary qualities—the same mechanical attributes, plus texture, invoked by Boyle—from '*secondary Qualities*' that *also* reside within the object."[41] In other words, in spite of his corpuscular philosophy, Boyle nevertheless maintains a separation between things and persons, while Locke altogether refuses such a separation. As Thompson goes on to observe, "Locke's secondary qualities are effectively in things and in persons at once."[42] Furthermore, "qualities are elicited from bodies [only] by relations to other bodies—including, of course, the human body."[43] Thus, in Locke as in Boyle, empirical observation is only possible because of the material relationality between the observer and observed and, thus, is fundamentally reliant on a kind of material *im*modesty.

Building on the moral philosophy of John Locke, David Hume similarly recognizes the importance of such relationality. In fact, he emphasizes it more overtly as he extends its influence beyond empirical observation to the very constitution of the self as well as to the operations of reason. Jay L. Garfield's book *The Concealed Influence of Custom: Hume's* Treatise *from the Inside Out* offers an illuminating depiction of the centrality of material relationality within Hume's moral philosophy. Far from arguing for detached and neutral empiricism, Hume asserts the fundamental instrumentality of "imagination" to empirical observation. To quote Garfield, according to Hume, "we presuppose a causal link between the object and our senses."[44] In other words, we believe characteristics of an object *cause* our perceptions of the object. However, as Garfield proceeds to explain, "Hume observes . . . that we have no theoretical arguments to the reality of causal necessity, although we presume this necessity in our ordinary life and in doing science. . . . A few pages later, Hume argues that our tendency to reason causally is grounded neither in the senses nor in memory, but in the imagination."[45] Causality and, thus, what we believe we empirically observe is "a propensity of the mind, mediated by processes in the imagination and by social practices of explanation."[46] In fact, according to Hume's Pyrrhonic skepticism, we cannot reason with certainty that an external world even exists much less presume that we may gain direct and therefore neutral empirical access to it.[47] Moreover, Hume asserts, "All our reasonings concerning causes and effects are deriv'd from nothing but custom; and . . . belief is more properly an act of the *sensitive*, than of the cogitative part of our natures."[48] As Garfield explains, for Hume, "reason itself has no independent warrant as an instrument for gaining knowledge."[49] Thus, even our most abstract beliefs and reasonings are derived from our sensitive responses to material stimuli. Hume's moral philosophy is therefore fundamentally incompatible with the gentlemanly modesty that Bacon and his followers promote because there is simply no psychological experience that is unmingled with or disentangled from either the matter of one's experience or one's physical environment. In Hume as in Locke, Baconian modesty can only ever be a fantasy.

Furthermore, as Garfield shows, Hume repeatedly, and in various contexts, describes the individual's material openness and responsiveness to their surroundings and the others with whom they come in contact. According to Hume, one's very self is constituted only in relation to other selves. As Garfield observes, Hume maintains that "[w]e are . . . social animals, tuned to resonate to each other . . . . We therefore develop as persons in response to social engagement. . . . Our own view of ourselves is determined by the views of others, and the self we represent is an essentially social self."[50] Moreover, sympathy between humans is a "natural" and "autonomic" phenomenon—a "psychic force, akin to gravitation in the physical world."[51] Indebted to Sir Isaac Newton, Hume's theory of sympathy is founded *not* on a "theory of reasoning" but, rather, on a "theory of forces" that explains how "[t]he communication of psychic energy . . . allows a kind of expansion of our self to *include others*, and the attitudes we have toward ourselves to be extended to those others as objects. . . . Other persons are not simply objects among other objects: they (and perhaps other animals by extension or analogy) are so intimately related to us that their very identities overlap ours."[52] Not only is the Humean self intimately (and materially) connected to other selves. It is in fact *inclusive* of—and thus fundamentally intermingled with—other selves. As a consequence, we cannot help but care for others.[53]

Considering how prominent early modern British philosophers such as Boyle, Locke, and Hume recognized the reality and significance of material entanglement, what are we to make of the philosophical modesty that the founder of the new science, Francis Bacon, promoted and to which many of his philosophers aspired? Both Thompson's and Garfield's studies certainly invite a reevaluation of Enlightenment era philosophical modesty and the buffering of the self it implies. In fact, Thompson argues that her book "refutes literary history's dominant account of early modern science"—in particularly, this account's "literal portrait of empirical knowledge, which places science and representation in a phenomenal enclosure delimited by visible things."[54] *Sensitive Witnesses* also complicates this portrait as it shows how British women authors of the Enlightenment era used contemporary concepts of material entanglement and relationality to claim philosophical authority for themselves in spite of the fact that women's capacity to contribute to collectively produced and accepted philosophical knowledge was rarely taken seriously. However, whereas Thompson appears to dismiss the practice of early modern modest witnessing altogether, portraying it as a misrepresentation of the new science by twentieth- and twenty-first-century historians, I take a more moderate view. While I agree with Thompson's characterization of how much Enlightenment era philosophy and literature preclude the practice of modest witnessing, I would assert that it is nevertheless important to recognize how philosophers regularly aspired to and performed gentlemanly modesty in their public presentations and publications, even (and perhaps especially) when such modesty conflicted with the material entanglement and relationality that their philosophical practices and principles emphasized.

Though Boyle's corpuscular chemistry recognizes material relationality (including between the scientist and his specimens), the "naked" style of his published writings nevertheless performs, and thus conveys a sense of, a separation between himself and his objects of study. So too does his employment of experimental technologies, which effectively place him at a literal remove from the matter of his investigations. Thus, though I find Thompson's study of the material relationality implied by early modern corpuscular philosophy wholly convincing, I nevertheless find it compatible with Donna Haraway's portrayal of modest witnessing as a means by which to ensure philosophical authority. Indeed, as Haraway argues, gentlemanly modesty allowed male empirical philosophers like Boyle to avoid appearing to be "polluted by the body," to "guarantee . . . the clarity and purity of objects" and, therefore, to ensure their own ability to make reliable philosophical claims, in spite of their own material entanglement which they profoundly perceived.[55] To witness is not only to see but also to document how one sees.[56] I would argue that while the early modern philosopher's perception itself—including their sense of their own relationship to the objects that they perceived—may not have been modest, the written evidence they provided of such acts of perception (for instance, in the *Philosophical Transactions of the Royal Society* and other publications similarly intended to be read by their peers) nevertheless tended to be so. Thus, early modern philosophers cultivated what may be seen as a semblance of detachment that downplayed and even repressed their acutely felt sense of their own material entanglement. Their empirical practice itself may have relied on sensitivity, but their public performance of their empiricism rarely broadcast this. See for instance this passage from Robert Hooke's *Micrographia* (1665): "Nor, is there a less admirable and wonderfull *Mechanism* in the foot of a Spider, whereby he is able to spin, weave, and climb, or run on his curious transparent clew."[57] Here, even as Hooke acknowledges the astonishment and pleasure afforded by his microscopic investigation, he retains a reserved air that prevents such acknowledgment from violating the distance between the philosopher and his object of study. Were he to have exclaimed with wonder, and thus evoked his own sensitivity, he would have compromised his philosophic remove. Instead, he describes his wonderment through sober statement, thus perpetuating the spider's otherness as well as his own distance, control, and even superiority.[58] Such performed restraint persisted in published philosophical works intended for virtuosi audiences throughout the long eighteenth century.

As it thus persisted, such restraint helped ensure the continued exclusion of women from serious philosophical activities. Though Al Coppola is right to recognize that "the immodest and interested observer" embodied by "the virtuosa" took center stage in popular science at mid-century, he neglects to recognize key differences between philosophical works intended for broad audiences and those intended for narrower ones comprised of serious practitioners.[59] Consequently, I would argue that he extends his argument too far when he contends that "[t]he eager spectator," which "derives from the figure of the female enthusiast," would "come . . . to organize [British] Enlightenment natural philosophy" in the second half of the eighteenth century.[60] In the chapter where Coppola presents this argument, the only work he substantially addresses that is not primarily designed for a popular audience is Joseph Priestley's *The History and Present State of Electricity, with Original Experiments*. Regarding this work, Coppola argues,

> Priestley imagines "the present state of electricity" . . . to be a vast and majestic terra incognita on the cusp of discovery, a sublime landscape of natural knowledge that cannot be charted solely with the sober empiricism of a few highly learned experts. Rather, the work positively requires the unschooled exertions of enthusiastic amateurs who are prepared to throw themselves into a highly sentimental experimental practice and feel their way around, by sheer instinct and nerve, to heretofore unknown truths.[61]

However, Coppola neglects to recognize that Priestley expressly addresses this work *both* to a popular audience ("persons who have a taste for Natural Philosophy in general") and to one comprised of serious virtuosi ("electricians in particular") and that his celebration of the "sublime" pleasures afforded by the study of electricity is more directed at the former than at the latter.[62] Moreover, though Priestley does repeatedly celebrate the "sublime" pleasures afforded by the study of electricity (not only in his preface but throughout the work), when he proceeds to document his experiments with electricity in the main body of the work, Priestley nevertheless performs precisely the same kind of gentlemanly philosophical modesty that seventeenth-century empirical philosophers promoted and performed.[63] See for instance this introduction to his "Experiments on Excitation":

> Finding by my own experiments, and those of others, that a glass tube, out of which the air was exhausted, discovered no sign of electricity outwards, but that all its effects were observed on the inside; I imagined that, if the air was condensed in the tube, it would operate more strongly on the outside; so that an additional atmosphere would give it a double virtue. But the result was the very reverse of my expectations.[64]

Without any points of comparison, it is impossible to be convinced by Coppola's analysis that Priestley's *History* is indicative of a broad trend. However, even if we presume that it is thus broadly indicative, I would assert that Priestley's consistent performance of gentlemanly philosophical modesty throughout the body of the work reveals Coppola's claim that the "entertainment promised by Priestley . . . is the very engine of discovery in the science of electricity, the epistemological ground of the later eighteenth-century's sentimental empiricism" to be inaccurate.[65]

In actuality, at least within the British context, gentlemanly modesty persisted throughout the Enlightenment era as a philosophical convention and a prerequisite for philosophical authority.[66] Contrary to the impression that Coppola gives when he asserts that "the masculine modest witness of science ceded primacy to a new feminized (and, increasingly, female) figure," women continued to be excluded from natural philosophy that was supported, condoned, and publicized by and for the Royal Society and its members. A woman would not claim a place in the *Philosophical Transactions of the Royal Society*, the society's journal, until 1787 when Caroline Herschel presented "An Account of a New Comet" addressed to "Charles Blagden, M.D. Sec. R. S." Even then, Caroline Herschel was an exception to the generally accepted rule of excluding women from serious philosophical study and contribution. Her "Account" illustrates her perception of herself as an outsider and the obligation she felt, as a result, to assure readers of her humility while subordinating herself and her philosophical practice to her brother William Herschel, a prominent fellow of the Royal Society. Presented in the form of a letter to the then Secretary of the Royal Society, Charles Blagden, her account begins thus: "Sir, In consequence of the friendship which I know to exist between you and my Brother, I venture to trouble you in his absence with the following imperfect account of a comet."[67] As this opening sentence makes clear, Caroline Herschel is well aware that it is only because of her close relationship with her brother that her comet sighting is accepted into the *Philosophical Transactions*. That a letter from her brother William is included in the volume directly after her own, vouching for the authority of her sighting, only further demonstrates her philosophical dependence on him.[68] To help secure and maintain her philosophical authority, Caroline Herschel performs masculine modesty as she records her observations: "August 1, 1786, 9 h. 50', the object in the center is like a star out of focus, while the rest are perfectly distinct, and I suspect it to be a comet. Tab. 1. fig. 1 10 h. 33', fig. 2. the suspected comet makes now a perfect isosceles triangle with the two stars *a* and *b*."[69] However, because she is a woman, her philosophical modesty and, thus, her right to participate in this publication by and for male virtuosi are reliant on her brother.

It is in stark contrast to the conventional modest witness, who performs a sense of detachment even when their philosophical experiences or principles belie it, that the authors explored in the present study conduct themselves. As sensitive witnesses, these authors eschew the aloofness and sobriety of the modest witness and instead allow themselves not only to feel but also to express the profound effect that objects have on them and their empirical investigations. In the words of William Smith's English translation of Longinus's first-century foundational treatise on the sublime, *Peri Hupsous*, they allow themselves to be "sensibly affected" such that they "swell . . . in transport."[70] Far from aloof, these sensitive witnesses maintain a sense of their own psychological and material porosity and, thus, a recognition of the interpenetration between subject and object, observer and observed. Such a recognition further contributes to the sublime dimension of their knowledge-making as it entails a breaking down and even, at its most extreme, a complete dissolution of the isolated self. According to the most prominent theorist of the sublime in Enlightenment England, Edmund Burke, we experience the sublime "when we contemplate the Deity," for when we find ourselves so doing, "we shrink into the minuteness of our own nature, and are, in a manner, *annihilated* before him."[71] In the cases of perception and observation that *Sensitive Witnesses* investigates, such annihilation of self occurs not only in "contemplation of the Deity" but in contemplation of the material world, including its most trivial and mundane elements. Softened, penetrated, charmed, seduced, intrigued, devastated, or emotionally moved by that which they witness, these sensitive witnesses open themselves to and commune with the beings and objects that inhabit their environments. As they do so, they suggest that knowledge is best made when an ontology of individuality (or individualism) is replaced by an ontology of kinship according to which self and other are continuous and, thus, united with rather than alienated from one another.

Beginning in the 1640s with Lucy Hutchinson's translation of Lucretius's Epicurean poem *De Rerum Natura* and concluding with Charlotte Smith's 1807 epic poem *Beachy Head*, *Sensitive Witnesses* traces and theorizes the practice of sensitive witnessing across Britain's long Enlightenment. As it does so, it appeals for a reconsideration of what the Enlightenment entailed and, thus, represents. For several decades, critical theorists have regularly criticized the Enlightenment era, portraying it as having provided the foundation for modern man's tendency to oppress, exploit, and dominate Others, both human and nonhuman. It has become commonplace to identify the Enlightenment era as the beginning of the Anthropocene, a new epoch in which human activity has become the prime mover of ecological change. James Watt's development of the steam engine in the 1780s is often cited as an important early touchstone in the epoch's historical emergence.[72] According to this analysis, climate change and other forms of ecological devastation that are becoming increasingly difficult to ignore as their effects grow increasingly apparent are by-products of the ambition and arrogance that the Enlightenment fomented.

While I recognize the persuasiveness of such analysis, with *Sensitive Witnesses* I refine the story that such critiques of the Enlightenment imply. As Aaron Hanlon has argued, "perhaps one of the most formidable challenges of the Anthropocene age is to rethink the kind of 'Real Knowledge' that positions nature as an inanimate object in relation to the human subject, and thus to learn how to articulate our world as animated with us, not in opposition to us."[73] This book rethinks such "Real Knowledge" by telling a story of the Enlightenment in which, through sublime knowledge-making and sensitive witnessing, philosophers develop an appreciation for their continuity with the material world—an appreciation that sets the foundation for a new sense of ecological responsibility and, ultimately, a new commitment to an ethic of care. This story does not cancel out more conventional depictions of the era of Enlightenment but, rather, disputes their claim to comprehensiveness. *Sensitive Witnesses* calls for a more inclusive approach to the Enlightenment and its legacy by uncovering philosophical practices and affective experiences that challenged the monolithic status of Baconian philosophy in Enlightenment era Britain. It helps not only to expand our historical understanding but also to broaden how we understand and experience what it means to be human in the modern world. I share the goal Lynn Festa outlines in her recent book *Fiction without Humanity: Person, Animal, Thing in Early Enlightenment Literature and Culture* to emphasize "the insecure foundation of humanity" while simultaneously recognizing that the distinguishing of man from matter (people, creatures, and things that are not man) was essential to the dawning of modernity.[74] *Sensitive Witnesses* shows that women found particular ways of engaging with, and even capitalizing on, such insecurity. Thus, I argue that their writings are particularly well equipped to, quoting Festa, "help us reconceive the rights to which we—whoever we are—seek to lay claim."

In her recent book, *Sweet Science*, Amanda Jo Goldstein demonstrates how certain authors of the Romantic period employed Lucretian materialism to legitimate their claim to philosophical authority and knowledge, thereby producing "an alternative . . . romanticism that runs athwart its usual basis in the English reception of Kantian and post-Kantian idealist philosophy."[75] *Sensitive Witnesses* complicates the originality of the Romantic Lucretian materialism that Goldstein highlights as it demonstrates that a number of seventeenth- and eighteenth-century British female authors had already, prior to the Romantic authors that Goldstein studies (Blake, Goethe, and Shelley), pursued and promoted a "sensitive" and thus "sweet" science, using poetry and other forms of literary expression as "privileged" sites of philosophical investigation and observation.[76] Goldstein contends, "Romantic, revisionary poetic sciences . . . challenged emergent life-scientific *and* aesthetic protocols to understand 'raw' sensation itself as susceptible and generative of social and rhetorical transformation; to countenance the mutual, material influence between the subjects and objects of experiment; and to position vulnerability—to impression, influence, and decay—as central, not inimical, to biological life."[77] While such Romantic "poetic sciences" may have been "revisionary" of what had become conventional scientific practices and principles by the turn of the century, they nevertheless hearken back to the earlier literary-philosophical productions that the present study investigates. Though scholars have, over the past several decades, begun to appreciate the influence of eighteenth-century writers on Romanticism, it is still all too common to take Romantics at their word when they claim to have made a break with the Enlightenment and, consequently, to overlook their indebtedness to earlier authors. This is particularly true of the influence eighteenth-century female authors had on canonical (male) Romanticism.[78] Considered in relation to Goldstein's *Sweet Science, Sensitive Witnesses* suggests that there is a connection to be made between feminist materialism of the British Enlightenment and British Romantic materialism. Even if the authors studied here did not have a direct influence on the Romantic authors that feature in Goldstein's book, their similarities are so striking and significant that they certainly have the potential to open up new possibilities of understanding the complex relationship between Romanticism and the Enlightenment.

In chapter 1, I explore how Lucy Hutchinson and Margaret Cavendish develop a practice of sensitive witnessing from opposite sides of the political spectrum that, directly influenced by Lucretian Epicureanism, sets an important foundation for transforming one's association with matter, as a woman, from an encumbrance into an asset. Insisting that all beings (whether male or female) are prone to the pull of and identification with matter, these author-philosophers challenged Bacon's investment in the modest witness. They instead promote an empirical practice that values the physical and psychological intimacy facilitated by atoms and that uses such intimacy for philosophical discovery and insight. In chapter 2, I explore how, like Hutchinson and Cavendish, Aphra Behn drew significant inspiration from Lucretius's *De Rerum Natura*. In this chapter, I assert that Behn presents a strong critique of contemporary philosophical practice while proposing an alternative that is distinguished by libertine Epicureanism. Behn encourages her readers to appreciate how sensory stimuli move and beguile our will, making us profoundly vulnerable to our environments. While such vulnerability threatens Baconian empiricism, which is invested in masculine modesty, it proves to be a boon to the philosophical practice that Behn models and promotes. It enables discoveries and insights that are unavailable to the masculine modest witness and, in so doing, shows that femininity is something to be proud of as a philosopher. Behn parodies masculine philosophical modesty and sets a foundation for the appreciation of feminine sensitive witnessing.

In chapter 3, I turn to the work of Eliza Haywood. I argue that Haywood presents a sensitive witness who is distinguished by skeptical doubt. Philosophers should, according to Haywood, practice sensitivity so that they may maximize their access to empirical truths. However, they should take care not to become swept up by their observations so that they may avoid becoming dogmatic or overly attached to a certain result of their enquiries. They must remain steadfast and thus resilient in the face of whatever truths come their way. Haywood directly challenges the modest witness and, specifically, its efficacy and ethics. She both legitimates and emphasizes the value and importance of feminist empirical philosophy. In the final chapter, I show how, at the end of the Enlightenment era, Charlotte Smith preserved the philosophical value of sensitive witnessing even as it became associated with the "cult of sensibility," which had become a target of satire and criticism by the end of the century. For Smith, the sensitive witness must combine feeling and reason. In doing so, they not only avoid the superficiality that was regularly seen to characterize sensibility but also cultivate an ontology of kinship and an ethic of care, which are, according to Smith, valuable to both philosophy and the public good.

The earliest author-philosophers I study here acquired support and inspiration for their innovative views and practices from Lucretius's Epicurean epic poem, *De Rerum Natura*. In recent years, scholars have become increasingly interested in Lucretius's influence on Enlightenment philosophy and literature. There is a general consensus now that his popularity in the period helped rouse an Enlightenment "neo-Epicurean revival."[79] Two main themes of *De Rerum Natura* are atomic interconnectedness and the material nature of humanity. Following Epicurus, Lucretius denies man's separability from nonhuman nature and portrays human as well as nonhuman existence as wholly determined by atomic interactions. The sensitive witnesses that I explore use such materialist principles to challenge the feasibility of modest witnessing and to lend support to their own distinctively affective form of knowledge-making. While Lucretius's Epicurean materialism directly informs the philosophical practice of the first three authors in this study—Hutchinson, Cavendish, and Behn—the final two authors, Haywood and Smith, espouse a materialism that is not explicitly identified with Lucretius or Epicurus. Theirs is instead symptomatic of how, by the middle of the eighteenth century, materialist concepts such as atomic transcorporeality, physical continuity, and the corporeality of thoughts and feelings had become so ubiquitous in England that one could invoke these concepts without attributing them to a particular philosopher or philosophy. *Sensitive Witnesses* shows that these various prominent female philosophers of the British Enlightenment chose feminine sensitivity over masculine modesty because they believed that it was both more sustainable and more capable of enhancing their ability to make reliable and illuminating observations about our atomically charged material universe. If man is inseparable from nonhuman nature, then claims to purity and transcendence are spurious at best and certainly not as authoritative as the modest virtuoso would have them seem.

The female philosophers *Sensitive Witnesses* addresses were not essentialists. They did not believe that women were intrinsically closer to matter than men. Rather, they understood such feminine closeness as a social construction and one of the many ways in which patriarchal society disadvantaged women. By developing and promoting sensitive witnessing, though, they were able to transform their close identification with matter from a symptom of their inferiority into an asset and even an indication of their philosophical superiority. While the modest witness conducts philosophy by the clear lights of indifferent observation and cool reason, the sensitive witness does so by the vibrant lights of embodied perception and engaged reason. In the face of matter, the sensitive witness welcomes physical and emotional stimulation whereas the modest witness is intent on remaining sober. Indeed, for the sensitive witness, the best knowledge-making is informed by experiences in which the very boundaries of the self are breached.

For each of the author-philosophers studied here, knowledge-making was both a philosophical endeavor and a sublime experience. They encourage nature's touch, both physically and emotionally. As they do so, they submit to the realization that they are themselves material beings and thus, in spite of their perception of themselves as individuals, they recognize their own continuity with their environments. To do so is a discomposing, often transporting, and sometimes annihilating experience. In his book, *A Secular Age*, Charles Taylor depicts the modern self as distinguished by a capacity to "tak[e] a distance from" or to "disengage[e] from everything outside the mind." According to Taylor, the modern self is fundamentally "buffered."[80] The pioneering women featured in this book, however, resist becoming such a self. Far from "buffered," they experience themselves instead as transcorporeally entwined with, and thus physically and emotionally joined to, their environments. Longinus argues that the sublime, "endued with Strength irresistible, strikes home, and triumphs over every hearer."[81] Though Longinus was writing in reference, specifically, to sublime rhetoric, in the Enlightenment era, his definition of the sublime was applied to the natural world as well. Eighteenth-century theorists of the sublime, such as Edmund Burke and Immanuel Kant, maintained that, for nature to be sublime, one had to be at a safe distance from it. Thus, they devised a masculine modesty for aesthetic experience that closely corresponded to that devised by Bacon for the new science. The philosopher-authors *Sensitive Witnesses* explores fundamentally challenge modesty both aesthetic and empirical as they develop a practice of sensitive witnessing that is not only philosophically illuminating but also generative of the sublime. For these pioneering women, the material world is a source of sublimity, but rather than standing at a distance from it so that they may admire it, they subject themselves both physically and emotionally to its "Strength irresistible" while recognizing themselves as a part of it. Thus, for these women, the sublime "strikes home" in a more dramatic, all-consuming way than Burke and Kant were ever willing to allow.

As many historians of science and ecocritics have shown, Bacon's distinctive variety of modest witnessing appeared to legitimate grand projects of environmental manipulation and exploitation.[82] Baconians may have performed modesty, but their deepest ambitions and most far-reaching projects were anything but humble. In *Modest_Witness@Second_Millenium.FemaleMan_Meets_OncoMouse* (1997), Donna Haraway proposes to "queer" Bacon's modest witness so that we might "enable a more corporeal, inflected, and optically dense, if less elegant, kind of modest witness to matters of fact to emerge in the worlds of technoscience."[83] I would like to conclude this introduction by suggesting that the author-philosophers *Sensitive Witnesses* studies should be understood as early modern precursors to the "queer modest witness" that Haraway proposes. As Haraway explains, Bacon's modest witness boasts of both neutrality and transcendence.[84] Moreover, empirical modesty is closely tied to and informs the association between technoscience and masculinity—an association whose legacy unfortunately persists in our present day, as the persistent underrepresentation of women in the hard sciences attests. Haraway seeks to queer the modest witness to achieve both a more inclusive science as well as a more ethical one that not only acknowledges but is fundamentally inflected by humanity's kinship with nonhuman nature.

Whereas Bacon's modest witness transcends nature, Haraway's queer variety remains interwoven with nature. As a direct consequence of this radical intermingling, she experiences the creatures, both sentient and nonsentient, with whom she inhabits the earth as "kin." For Bacon's modest witness, such creatures are fundamentally Other than himself. In contrast, Haraway's queer modest witness "becomes with" her fellow creatures.[85] She sees herself as bound up with them: "one must be in the action, be finite and dirty, not transcendent and clean."[86] Bacon's method fosters an ontology of detachment; Haraway's, an ontology of kinship. As a result, whereas Bacon's modest witness is perfectly willing to sacrifice life for knowledge, Haraway's queer witness is much more circumspect. Bacon encourages an ethic of discovery and advancement; Haraway an ethic of care. Bacon's modest witness pursues progress for its own sake, without much if any concern for the sacrifices it exacts. Haraway's queer witness's pursuit of progress is always accompanied by a deeply felt ecological responsibility. All of these features of Haraway's queer witness may be detected within the Enlightenment author-philosophers that *Sensitive Witnesses* investigates. Thus, as we seek to queer scientific practice in the twenty-first century while we reckon with the increasing ecological catastrophes that our unabashed consumption and exploitation have created, I would propose that we take inspiration from the pioneering women of the Enlightenment era who refused to conform to masculine modesty even as they sought to assert themselves both as philosophers and as authors of the sublime. Because of the prominence of Baconian empiricism as well as aesthetic philosophies, such as those of Burke and Kant, that similarly recommend masculine modesty—distance, detachment, sobriety, aloofness—nature has become perceived as "a reified thing in the distance."[87] However, *Sensitive Witnesses* shows that there was a significant feminist tradition of not only critiquing such modesty but also offering alternatives to it in the Enlightenment era. I share this study in the hopes that we may together learn about these pioneering women but also so that we may take inspiration from them as we seek to heal and preserve our material world.

ONE

EPICUREAN SOFT SCIENCE: LUCY HUTCHINSON AND MARGARET CAVENDISH

And now this *Iron Age*'s so *rusty* grown,
That all the *Hearts* are turn'd to hard *flint-stone.*
—MARGARET CAVENDISH, "Of the death and buriall of Truth"[1]

WHY WAS THE FIRST English translator of Lucretius's Epicurean masterpiece *De Rerum Natura* a woman? According to her own recollection, Lucy Hutchinson produced her translation alongside her needlework while her children were taught by their tutors in the same room. She explains that she embarked on the translation "out of youthfull curiositie, to understand things I heard so much discourse of at second hand."[2] However, translating such a long and complex poem was no casual task, and its atheism only accentuated the seriousness of the undertaking. It is therefore hard to see a general curiosity as, in itself, a sufficient rationale. As David Norbrook posits, the project "demands a commitment that would surely be easily sapped by complete hatred of its ideas."[3] According to Norbrook, the translation may have been an attempt to compete with her adversary, the eccentric philosopher Margaret Cavendish, whose *Poems, and Fancies* (1653) was deeply indebted to Lucretius.[4] Or, it may have provided a framework for her to pursue a regular Latin practice.[5] Norbrook argues, though, that "there were elements in the poem that could justifiably appeal to her."[6] This chapter builds on the scholarship of Norbrook and others to deepen our understanding of Lucretius's appeal to

Hutchinson. Analyzing her Lucretius alongside her great Biblical epic *Order and Disorder* (1679), I argue that Lucretius provided Hutchinson with a foundation for an alternative variety of knowledge-making, more readily accessible to women than the conventional and thus more familiar version that was simultaneously being developed and promoted by the Royal Society.

Francis Bacon pioneered a distinctive philosophic orientation to nonhuman nature that perceived man as both separate from and dominant over the material world. He therefore encouraged the scientist to be a "modest witness."[7] Cultural historians have rightly acknowledged the profound influence of Bacon's modest witness, and the method of psychological purification it entailed, on early modern English empirical practice. However, in focusing so heavily on the modest witness, we have neglected to recognize the significance of a strikingly different type of witness that emerged at around the same time—what I call the *sensitive witness.* It is through sensitive witnessing that Hutchinson develops her alternative form of knowledge-making, but Hutchinson is not alone. Her contemporary Margaret Cavendish similarly uses sensitive witnessing to devise an alternative philosophical method of discovery. Both female philosophers root their authority not in a sense of their autonomy from the material world as Bacon does but, rather, as they recognize the "trans-corporeality" of everyday life, in an appreciation of their kinship with it.[8]

Bacon's Modest Witness

In his *Novum Organum*, Bacon outlined how to cultivate a philosophic mind that would be capable of pursuing the innovative empirical method he sought to promote. Not just any frame of mind would do. He indicated that the success of his method was contingent on the scientist's success at triumphing over what he called "the idols of the mind."[9] To achieve such success, the scientist would have to aspire to a mind "free and cleansed" and, thus, detached from as well as transcendent above the material world.[10] This variety of philosophic mind has informed scholars' theorization of the early modern "modest witness." A primary way in which the Baconian empiricist maintained his "freedom" and thus detachment from the matter of his investigations was through the cultivation of modesty. By engaging in modest witnessing, the scientist ensured that "his narratives could be credited as mirrors of reality."[11] The early modern scientist was therefore expected to keep himself out of his own experimental pursuits: he was to be "a drudge of greater industry than reason";

he was to write in a "plain, ascetic, unadorned" and "functional" style; and, he was to distinguish, in his writing, between "experimental findings" and "their interpretation."[12] In these ways, he would rise to the level of "a disinterested observer," his mind purified or, quoting Bacon, "expurgated" of the matter in which he immerses himself.[13] As the following observation shows, the "intermingling" of subject and object, man and matter, was a problem for Bacon: "And the human understanding is like a false mirror, which, receiving rays irregularly, distorts and discolours the nature of things by mingling its own nature with it."[14] Bacon's answer to this problem of intermingling was for scientists to prepare for their experiments by cultivating a philosophic mind that entailed a transcendence of, and thus separation from, matter. *Leviathan and the Air-Pump* demonstrates how influential and widely adopted Bacon's methods of psychological purification and modest witnessing became in the early decades of the new science. Moreover, other scholars such as Lorraine Daston, Peter Galison, Donna Haraway, and Tita Chico have recognized the enduring legacy of these methods.[15]

While the Royal Society may have periodically professed a "come one, come all" attitude, thus appearing to encourage wide participation in its empirical pursuits, its ban on women attending its meetings is an indicator of the exclusionary nature of its methods.[16] Distinguished by the practice of "modest witnessing," early English science associated reliable knowledge with the notion of a detached and contained self and did so even when its principles challenged the possibility of such detachment.[17] How were women to achieve the autonomy from matter required by Baconian modesty when they weren't even presumed to have control over their own personhood? As Donna Haraway explains, "The issue was whether women had the independent status to be modest witnesses, and they did not."[18] Feminine and masculine modesty have been historically perceived as binary opposites. The former is "of the body," enacted by women's preservation of their virginity; the latter is "of the mind," enacted by men's philosophical detachment from matter.[19] Consequently, "within the conventions of modest truth-telling, women might watch a [scientific] demonstration; they could not witness it."[20] From the outset, the Baconian modest witness was presumed to be a gentleman.

Lucy Hutchinson's Lucretius

That women were expected to manage the domestic operations of the household only further accentuated their inability to assert their autonomy from their material environments. The image of Lucy Hutchinson translating Lucretius alongside her needlework with her children engaged in their studies in the same room offers an emblematic illustration of women's close association with domestic materiality: "I turnd it into English in a roome where my children practizd the several qualities they were taught, with their Tutors, & I numbred the sillables of my translation by the threds of the canvas I wrought in, & sett them downe with a pen & inke that stood by me."[21] Surrounded by women's work, Hutchinson's writing space is hardly a space of her own and is far from philosophically autonomous from its material environment. Indeed, her depiction suggests that Hutchinson experiences her writing practice itself as a material activity—hardly different from, and in fact closely corresponding with, her embroidery. This passage highlights the physicality of her act of translation. "[S]ett downe with a pen & inke," it is as much a work of her body as it is of her mind. Even if Hutchinson had wanted to compose her translation at a distance from the matters of her household, the practical pressures they exerted on her would likely have prevented her from doing so.

According to Bacon's portrayal of the modest witness, the undeniability of her association with materiality would put the female philosopher at a significant disadvantage, even potentially precluding her from engaging in respectable philosophical practice altogether. However, Hutchinson chose instead to view such material association instead through the framework of the Epicureanism that Lucretius poeticized. As a result, her affinity with matter becomes something altogether different. Epicurean philosophy normalizes the irresistible pull of one's material environment and insists that all beings (whether female or male) are equally exposed to it. As Monte Johnson and Catherine Wilson assert, "[b]oth the atomic reality alleged to underlie the appearances and the self-sufficiency of nature forcefully asserted by Lucretius exercised a powerful influence on modern science, and his name was still being regularly invoked in scientific contexts as late as the nineteenth century, with his influence formally acknowledged well into the twentieth."[22] Hutchinson uses Lucretius's currency to her advantage. According to Epicurean materialism, "[A]ll nature . . . / Consisteth . . . of two parts, which be / The bodies and th'immense vacuitie."[23] Nature, in other words, makes no distinctions among beings. Whether male or female, human or animal, even sentient or nonsentient—all of the created world is composed of "bodies and vacuitie." Such continuity extends down to the atomic level: according to Epicureanism, all beings are composites, "confluence[s] of seeds" or atoms.[24] Moreover,

> motions doe
> Dissolve all creatures, and their frames renew;
> Whither first bodies, by impulsive force,
> Or moving faculties, maintaine their course,
> Wandring for ever in th'unbounded space.[25]

Atoms are in a constant state of transcorporeal movement, and this movement continually "renews" the "frames" of "all creatures." Comprised not only of atoms but also of "vacuitie" or void, all bodies are porous and, with the perpetual movement of atoms, are therefore perpetually penetrated by the atoms of other bodies. Thus, there can be no autonomous being in the Epicurean universe: "no fixt seate can be for bodies found."[26] All beings are inseparable from their environments. In fact, because of the fundamental transcorporeality of the created world, it becomes difficult to distinguish a being from its environment at all.

No wonder, then, that it was a desire "to understand things [she] heard so much discourse of at second hand" that motivated Hutchinson's translation of Lucretius.[27] Epicureanism provides Hutchinson with a rationale for her own access to knowledge that would otherwise remain out of reach to her. As a woman, she is incapable of becoming a true philosopher according to orthodox Baconian methods: she can never fully rise to the level of the modest witness and, consequently, can never adopt the requisite "naked way of writing."[28] Thus, according to the guidelines of Baconian empiricism, all philosophic knowledge can only ever really be "second hand" to her. However, Epicureanism challenges the legitimacy of Baconian modesty by denying the possibility of autonomy from matter.[29] With her translation of Lucretius, Hutchinson reveals that Baconian modesty can be nothing more than an illusion. It is therefore not only an understanding of Epicureanism that her translation of Lucretius affords her. Rather, it makes all of philosophy available to her, even giving her the authority necessary to observe and analyze as a philosopher herself.

Epicurean Materialism and Hutchinson's *Order and Disorder*

These philosophical principles of Epicureanism would provide an important foundation for Hutchinson's epic masterpiece, *Order and Disorder*, which she went on to publish in 1679, twenty or so years after she completed her translation of Lucretius. That Hutchinson appears to distance herself from *De Rerum Natura* in her preface to *Order and Disorder*, suggesting that Lucretius had, in her youth, "fill'd [her] brain with . . . foolish fancies" and "ugly wild impressions," is belied by her reliance, throughout *Order and Disorder*, on many of the terms and concepts that feature in his poem.[30] Recent scholarship has begun to illuminate intertextual affinities between Hutchinson's *De Rerum Natura* and *Order and Disorder*.[31] Alvin Snider argues, for instance, that Hutchinson's Biblical epic "borrows from Lucretius . . . in incorporating certain elements of its materialist account of perception and in its depiction of dreaming as a somatic state that reveals something crucial about the operations of the mind."[32] According to Snider, "[a] retelling of Genesis provided her with the means to engage contemporary materialism without abandoning an appearance of piety."[33] Mihoko Suzuki considers how Hutchinson's portrayal of the relationship between humans and animals in *Order and Disorder* is indebted to Lucretian materialism. She argues that Epicureanism significantly informs the "leveling" of "the accepted hierarchy of human over animal" that Hutchinson portrays in *Order and Disorder* as an important component of her broader commitment to democracy.[34] Exploring Hutchinson's indebtedness not only to Lucretius but also to the Calvinist minister John Owen, Cassandra Gorman reveals how the author's depictions of material existence in *Order and Disorder*—in particular, its propensity to disintegrate—echo Lucretius. According to Gorman, "Hutchinson reinterprets Lucretian imagery to shape [the] material and providential universe" of her Biblical epic. "The ultimate promise of her concentration on material disorder . . . is that soteriological order will be reconstructed from the ashes."[35] Considering such persistent traces of Lucretius in Hutchinson's writings, it is no longer possible to maintain that Epicurean materialism was merely a passing fancy for her. As Jonathan Goldberg observes, even Reid Barbour, who tends to emphasize the significance of Hutchinson's repudiation of Lucretius in her preface to *Order and Disorder*, indicates that her "connection to Lucretius . . . could not be undone by the prefatory letter repudiating the translation."[36] Rather, his Epicurean masterpiece continued to inform her views of the material world decades after she produced her translation of it.

I argue that it is in *Order and Disorder* that Hutchinson develops and promotes her philosophical method of sensitive witnessing, a method deeply indebted to Epicureanism. Hutchinson's is an embodied, intimate, and immersive philosophical practice whose strength resides not in its detachment from, but rather in its responsive openness to, matter. Indeed, according to Hutchinson, the philosopher should be willing to be profoundly affected and even transformed by matter—a process that she portrays as a kind of softening. According to Hutchinson's depiction, humans are all too prone to becoming "hardened" by "impunity."[37] Sinners that we are, we often behave like Cain, "Who, hardened in his pride and arrogance / Raised him a city with aspiring walls," designing works that express and thereby externalize the psychological rigidity that our hubris produces.[38] Naturally "rebellious," we too often "lose all sight and sense and motion, stiffened by impertinence."[39] Deeply critical of such rigidity, the narrator of *Order and Disorder* instructs that as long we remain "hard as rocks" we will persist in being "unmoved" by God and thus will stubbornly in our "own pernicious, lewd paths trod."[40] Hutchinson significantly draws on Epicurean materialism to depict the softening that she advocates as an alternative to such sinful rigidity. In her depiction of the crew aboard Noah's ark in *Order and Disorder*, she associates the sailors' "flinty hearts," made to "tremble" like "leaves in autumn," with "the firm stones" that have, in the deluge, been "shattered into atoms" by "sick Nature's groans."[41] Here, Hutchinson's reference to "atoms" evokes Epicurus, and her suggestion in this passage that man's atomistic material nature makes him prone to suppleness even when he appears rigid closely corresponds to passages from her translation of Lucretius. Within the Epicurean universe that Lucretius poeticizes and Hutchinson translates, matter appears in a constant state of flux. Indeed, we are barely able to assign solidity to bodies—"matter *flows*."[42] Moreover, their shapes are fleeting. Throughout *De Rerum Natura* we are confronted with the disintegration of matter. Erosion of hard objects is a recurring theme:

> [I]n revolution of some yeares
> The ring of mettall on the finger weares
> The drops of water harder stones decay,
> The iron of the plow time weares away.[43]

Even metal and stones are prone to being molded and transformed by external forces.

Bodies are soft, malleable, and entropic according to the Epicurean materialism that Lucretius expresses, and Hutchinson imports these qualities into the Biblical landscape of *Order and Disorder.* Lucretius's depiction of a "rolling flood" has particular resonance:

> But with no lesser force and fury goe
> The inundations which from mountains flow
> When store of raine the rappid torrent fills
> Whose violent streame descending the high hills
> Bears downe the groues, and vineyards, overthrows
> Bridges, whose vanquisht strength cannot oppose
> The suddaine furie of the waters fall,
> Which carries downe great stones, banks, rubbish all
> That in their passage lies; thus with loud noyse
> The rolling flood whatere it meets destroyes.[44]

Notice the close resemblance of this passage from her translation of *De Rerum Natura* to her depiction of the deluge in *Order and Disorder*:

> Prodigious thunders followed the dire blaze,
> Filling all hearts with terror and amaze
> While they the rocks, the oaks and mountains rent,
> Nor with less fury tore the firmament.
> These fragors thus unjointing the whole frame,
> Heav'n and Earth Chaos once again became. . . .
> And Heaven's upper windows, open thrown,
> As violently poured their waters down.
> While the rude tempest thus the high roofs shook,
> The crystal casements were to shivers broke
> And the world's solid walls like ice did melt
> When they the heats of divine vengeance felt.[45]

Both passages highlight the "violent" capacity of water to move mountains. Rushing "downe," it becomes a force of destruction, not only carrying with it debris along the way but also disintegrating large solid forms—in the former, "groues," "vineyards," and "bridges"; in the latter, "rocks, . . . oaks and mountains." Even the "crystal casements" of "Heaven's upper windows" are not immune to the dissolving force of the deluge. What were once "the world's solid walls" become "like ice," "melt[ing]" in "the heats of divine vengeance" of which the flood is an expression. Material solidity appears here much as it does in Lucretius's depiction as provisional if not completely illusory.

While Hutchinson acknowledges the tragedy of such material dissolution, associated as it is with the death of man and his works, she appears even more intent on emphasizing its redemptive capacity. As the human world becomes shaken, fragmented, and ultimately dissolved, it doesn't disappear but, rather, becomes integrated with God and is thus, through softening and dissolution, redeemed. Given that the flood waters are an expression of God's wrath, for the "solid walls" of "the world" to "melt" is for them to become united with God. Humanity's disintegration expresses a kind of baptism whereby humanity is purified, restored, and made open to God's grace.[46] All is not lost in such melting. On the contrary, it affords the opportunity for a dramatic salvation for humanity, whereby our souls "break," like Jacob's, "into a pious admiration," and we "convert" our "wild dread to reverence."[47] Indeed, "by [God's] grace [we] attain a blessèd state" and are thus "[r]edeemed out of the dark unpleasant grave."[48] In the biblical universe that Hutchinson depicts in *Order and Disorder*, such redemption is predicated on God's infinitude and his omnipresence, which she emphasizes through the voice of God in her narration of Jacob's dream vision: "Lo, I am with thee and in every place / Where'er thou goest will be thy guard; my grace / Shall thee in all thy wanderings sustain."[49] God's permanence bolsters humanity in the face of our entropic destruction, sustaining us even as we wander, whether motivated by our own will or impelled by the sometimes violent vicissitudes of God's material creation. As material beings, we are subject to malleability, mutability, and finitude. God, in contrast, is fixed and infinite, and as we submit to our own impermanence, we find ourselves capable of resting in His protective love and grace.

Given the atheism of Epicureanism, we may expect this sentiment of finding rest in the permanence of God to separate *Order and Disorder* from *De Rerum Natura*. However, even this dimension of Hutchinson's biblical epic evokes a key component of Lucretius's atheist one: namely, the reassurance that comes from recognizing the permanence of atoms. Matter may be seen to "flow" throughout *De Rerum Natura*, yet Lucretius insists that atoms ("first bodies") are "uncompounded," "solid in themselves," and "indued with immortality":

> But if first bodies without vacuums be
> And solid, they must haue eternitie

Besids unlesse eternitie had bene
In the first matter, we had long since seene
All things reduc'd to nothing, and soe had
Againe from nothing sprung, which we haue made
So much impossible we must conclude
The principles of all things are indued
With immorality, and the last day
Dissolues all creatures into them, that they
To the next births new matter may supplie
And still preserue their firme simplicities.[50]

Lucretius here depicts atoms to sustain material beings even as these beings dissolve. "Solid" and imbued with "eternitie," they regenerate life, "to . . . next births new matter . . . suppl[ying]." Such eternal sustenance and vital regeneration closely correspond to the life preserving and renewing power of God Hutchinson depicts in *Order and Disorder*. Indeed, Hutchinson acknowledges this connection herself when, in the latter poem, she refers to those who have "miscalled [God's] high help accident."[51] In a footnote to this line, David Norbrook directs the reader to see Hutchinson's "condemnation of materialists in the Lucretius preface: 'leaving all things here, to Accident and Chance, deniing that determinate wise Councell and Order of things they could not dive into, and deriding Heaven and Hell, Eternall Rewards and Punishments, as fictions.'"[52] Norbrook is right to characterize this line as critical of Epicurean materialists. However, it is important to acknowledge the precise nature of this criticism. The Epicurean mistake, according to Hutchinson, is one of terminology—"miscalling"—and in portraying it as such, she quietly acknowledges the *similarities* between the Epicurean and Christian universes at least as much as, if not more than, she marks their differences. What Lucretius portrays as chance in *De Rerum Natura* is renamed "providence" by Hutchinson in *Order and Disorder*.[53] However, in both cases, the fundamental nature of things is the same: matter persists because it is imbued with a fixed permanence that counteracts bodily mutability.

The equanimity that resting in God provides in *Order and Disorder* parallels what is known as "ataraxia" (tranquility or "the freedom of the soul from disturbance") in the philosophy of Epicurus.[54] "Piety" is, in Hutchinson's translation of *De Rerum Natura*, "to receive all chance with calmed brests."[55] As in *Order and Disorder*, a recognition of and submission to our own mortality are crucial to our achievement of such tranquility. The organizing principles of both poems may be radically different from one another: chance governs *De Rerum Natura*, God's providence governs *Order and Disorder.* Nevertheless, both poems similarly advise their readers to accept and surrender to the motions of the universe. By doing so, we become capable of maintaining peace of mind even in the face of the most dramatic of material upheavals. This tranquility is, in both Hutchinson's *De Rerum Natura* and also *Order and Disorder*, "the goal of a blessed life."[56] By letting go of our attachment to material firmness and fixity, softening in response to the difficult collisions with which life presents us, we free ourselves from the vicissitudes of matter. In the words of Epicurus, "a man who lives among immortal goods is in no respect like a mere mortal animal"—be these immortal goods rooted in the eternity of atoms, as is the case in Hutchinson's translation of Lucretius, or in the eternity of God, as is the case in her biblical epic.[57] Thus, in both poems, Hutchinson emphasizes the value of a soft and malleable orientation to matter, and as she does so, she offers a profound recognition of man's materiality and, therefore, his intimate kinship with and sensitivity to nonhuman nature. Informed by Epicurean materialism, she places her hope for England in its citizens' capacity for sensitive witnessing—observing and reflecting on nature with tenderness, receptivity, and an appreciation for man's immersion within and embodiment of it. In doing so, she envisions a version of witnessing that diverges from that promoted by Francis Bacon and his followers. Bacon called on philosophers to transcend nonhuman nature, to resist "intermingling" with matter. In contrast, Hutchinson pleads with her reader to submit lovingly to their own transcorporeal material nature. To do so is not only to find tranquility in God's grace but also to become, as she teaches by her own example, a good philosopher. Thus, Hutchinson finds a different means by which to avoid the pitfalls of "the idols of the mind," and thus achieve philosophical authority, than Bacon proposes. Whereas Bacon seeks a purified philosophical mind afforded by a detachment from materiality, she seeks a tranquil philosophical mind that is afforded by a submission to materiality.

Though her portrayal of the value of sensitive witnessing reflects on contemporary natural philosophy, even implicitly presenting a critique of its methods, there is no evidence that Hutchinson was herself a natural philosopher. She admired her mother for her involvement with natural philosophy. In her autobiography, Hutchinson narrates how her mother, Lucy Apsley, had provided Sir Walter Raleigh and Patrick Ruthven, Earl of Gowrie, with the opportunity, while imprisoned in the Tower, "to make their rare experiments" in "chimistrie" "at her cost." According to Hutchinson, she did so "partly to comfort and divert the poore prisoners, and partly to gaine the knowledge of their experiments, and the medicines to helpe such poore people as were not able to seeke to phisitians. By these means she acquir'd a greate deale of skill, which was very profitable to many all her life."[58] Together with her daughter, Hutchinson supported her husband Colonel John Hutchinson's quasi-empirical hobby of "sorting and shadowing [i.e., drawing] cockle shells" by collecting them for him.[59] According to David Norbrook, the Hutchinson's "household at Owthorpe . . . [was] a centre for the display of fine works of art, ingenious landscaping, and cultivated recreation."[60] Empirical philosophy would have been at home in this environment. However, the readers she envisions both for her translation of Lucretius and for *Order and Disorder* are not, or at least not primarily, Baconian empiricists but, instead, distinguished by their religious affiliation—specifically, with Nonconformism.

When it was finally published in 1675, she dedicated her translation of Lucretius to Arthur Earl of Anglesey, a politician with Calvinist views. In her preface to *Order and Disorder*, she addresses herself to those who might "be as much affected and stirred up in the reading . . . to admire the glories and excellencies of our great Creator, to fall low before him, in the sense of our own vileness, and to adore his power, his wisdom, and his grace, in all his dealings with the children of men."[61] As this passage suggests, her primary goal in the epic is to inspire religious reverence and humility, and given her own Calvinism, we can presume she has fellow Nonconformists in mind. The subtitle of the poem reflects this sense of purpose: "The World Made and Undone. Being Meditations Upon the Creation and the Fall; As it is recorded in the beginning of Genesis." Of course, at the time of the poem's original publication in 1679, theology and natural philosophy tended to be viewed as closely related.[62] As the following observation made by William Derham during a discussion of the Boyle lectures illustrates, early English Protestantism and the new science mutually informed and reinforced one another: "As the noble Founder of the Lectures I have had the Honour of Preaching, was a great Improver of Natural Knowledge, so, in all probability, he did it out of a pious End, as well as in Pursuit of his Genius. For it was his settled Opinion, that nothing tended more to cultivate true Religion and Piety in a Man's Mind, than a thorough Skill in Philosophy."[63] Robert Boyle himself reinforces this sentiment when he writes in *The Second Part of the Christian Virtuoso*, "[T]he study of nature, with design to promote piety by our attainments, is useful, not only for other purposes, but to increase our knowledge, even of natural things, if not immediately, and presently, yet in time, and in the issue of affairs."[64] Boyle's lecture series was a product of his belief in the mutual dependency of Baconian empiricism and Christianity. Natural philosophy was widely perceived as "[n]ext to Gods Word . . . the most Soveraign Antidote to expel the poison of Superstition; and . . . the most approved food to nourish faith."[65] This perception helped inspire the physico-theology tradition, which found evidence for the existence of God through the empirical study of nature (particularly, the elements of nature that were seen to embody "design").[66] The physico-theology tradition, which Boyle's lecture series helped promote, was an influential outgrowth of this perception.[67] In fact, it was precisely as a means of celebrating God that Bacon first presented his empirical program, which would serve as the main experimental template for the Royal Society. In his plan of the *Great Instauration*, he includes the following prayer: "[M]ay [God] graciously grant to us to write an apocalypse or true vision of the footsteps of the Creator imprinted on his creatures. Therefore do thou, O Father, who gavest the visible light as the first fruits of creation, and didst breathe into the face of man the intellectual light as the crown and consummation thereof, guard and protect this work, which coming from thy goodness returneth to thy glory."[68] By empirically studying and, thus, reading nature, philosophers could, in the words of Nehemiah Grew, hear "GOD read . . . the World his own Definition, and their Duty to him."[69] Examining nature, they could bask in the glory of God. The approach to the natural world that Hutchinson takes in *Order and Disorder* closely corresponds to this method, as her careful observation of nature's details prompts her and, thus, her reader to reflect on God's goodness and grace. The method of sensitive witnessing she advocates is therefore relevant to contemporary Baconian empiricism.

Margaret Cavendish as Materialist and Sensitive Witness

However, because Lucy Hutchinson did not herself participate in empirical investigations, its relevance to the new science remains implicit. In the works of another early English female philosopher Margaret Cavendish, though, sensitive witnessing becomes translated into actual empirical practice. Throughout the rest of this chapter, I analyze the empirical implications of passages from several of Cavendish's works: her collection of verse *Poems, and Fancies* (1653), her miscellany *The World's Olio* (1655), and her major scientific works *Obser-*

vations upon Experimental Philosophy (1666) and *Grounds of Natural Philosophy* (1668). I argue that, inspired by Epicureanism and over the course of her career, Cavendish advocates a variety of sensitive witnessing that corresponds in striking ways to Hutchinson's. Though there is much that distinguishes these two philosophers from one another both politically and philosophically, the frames of mind and practices of observation that they advocate and model through their own examples are quite similar to one another. Taken together, they present a compelling critique of and alternative to the modest witness promoted by Bacon. Various scholars have pursued connections between Hutchinson and Cavendish.[70] Even Virginia Woolf associated the two authors when she commented in one of her essays, "I must drop my book, the pious Mrs Hutchinson, and leave her to make what terms she can with Margaret, Duchess of Newcastle."[71] David Norbrook argues that "[t]here was . . . an important dialectic" between them: "Hutchinson was arguably very powerfully influenced by Cavendish; it is less easy to show influence the other way but Cavendish was at least reacting to cultural forces with which Hutchinson identified."[72] Their political ideologies could hardly have been further apart, with Hutchinson a staunch republican and Cavendish a similarly staunch royalist. However, I would argue that, nevertheless, their philosophical methods coincided.

The only woman to have been allowed to visit the exclusively male Royal Society until centuries after its inception, Margaret Cavendish styled herself throughout her several published writings as a vocal critic of the society members' common beliefs and practices.[73] The precise targets of her criticisms and the exact characteristics of the system of nature that she envisions have been cause for debate among scholars. My analysis builds on that of Line Cottegnies and Emma Rees, both of whom demonstrate that, to quote Cottegnies, the "woman-friendly" dimension of Epicureanism provides Cavendish with a useful vehicle for asserting her own feminine philosophical as well as literary authority. Though some scholars argue that she ultimately abandoned materialism, I join Londa Schiebinger and Stephen Clucas in maintaining that materialist philosophies continued to influence her throughout her career.[74] Clucas offers a particularly in-depth, nuanced, and therefore persuasive analysis: "Margaret Cavendish's natural philosophical works are essentially atomistic, although we see in her theories a progressive attempt to refine and reconcile atomic ideas within a philosophical framework which opposes itself to the simple mechanism of 'classical atomism.'"[75] Though Clucas documents various ways in which, over the course of her career, Cavendish revises and refines the materialism that she espouses, he demonstrates that Epicureanism maintains a persistent appeal for her. Her publishing career begins with "a thorough-going Epicurean atomism," expressed in the first forty-five pages of *Poems, and Fancies*.[76] However, even when she publishes "A Condemning Treatise of Atomes" in *The Philosophical and Physical Opinions* of 1655, what she presents "is not an *anti*-atomist statement, but rather a refinement of the crudities of an updated Democritean-Lucretian atomism" that "retains a residual attachment to the broad principle of atomic structure. Her objections, it seems, are to the idea of *mechanical* atomism," not to atomism generally.[77] As Clucas explains, in her later works, "Cavendish's atomism is a synthesis of materialism and vitalism."[78] Much like Hutchinson, Cavendish began her career standing on the shoulders of Lucretius; then, as she matured, she developed a more complex relationship to him, though would nevertheless continue to take inspiration from vitalist materialism throughout her career even as she interrogated some of its central principles.

I would argue that, for Cavendish as for Hutchinson, vitalist materialism provides an important foundation for both her own philosophical and literary authority as well as for the philosophical methods that she would develop and promote.[79] As we will see, like Hutchinson, Cavendish is insistent on the continuity of matter and, informed by such continuity, denies man's ability to assert his autonomy from it. Indeed, Cavendish too maintains that the best observational and philosophical practices are those that acknowledge rather than deny material transcorporeality and, thus, humanity's openness, responsiveness, and sensitivity to its material environment. Cavendish presents serious criticisms of practices on which members of the Royal Society relied to affirm their "modesty."[80] Most notable of such practices are their use of optical instruments. As Tita Chico demonstrates, the microscopist Robert Hooke achieves Baconian modesty by making his body into an instrument, essentially objectifying witnessing by technologizing himself and thereby denying his own physicality.[81] Cavendish took a deeply skeptical view of microscopy. As she argues in a chapter from *Observations upon Experimental Philosophy* entitled "Of Micrography, and of Magnifying and Multiplying Glasses," the modest method of micrography is distorting rather than revealing: "magnifying, multiplying, and the like optic glasses, may, and do oftentimes present falsely the picture of an exterior object; I say, the picture, because it is not the real body of the object which the glass presents but the glass only figures or patterns out the picture presented in and by the glass, and there mistakes may easily be committed in taking copies from copies."[82] For Cavendish, far from ensuring a detached and, thus, neutral access to specimens, microscopes manipulate and therefore present a highly misleading and unreliable mediated depiction of them. She goes on to say, "Wherefore those that invented microscopes, and such like dioptrical glasses, at first, did, in my opinion, the world more injury than benefit; for this art has intoxicated so many men's brains, and wholly employed their thoughts and bodily actions about phenomena, or the exterior figures of objects, as all better arts and studies are laid aside."[83] Here, Cavendish presents a scathing critique not only of microscopy but also, implicitly, of Baconian modesty. As Eve Keller observes, Cavendish takes aim at Bacon and his followers' "claims of methodological rigor, value-neutrality, and objectivity" and perceived these qualities "not as monolithic conduits for achieving certainty, but as social constructions that are endorsed as much because they advance the needs of their adherents as because they are deemed to be scientifically effective or true."[84] Empirical philosophers presumed that their instruments helped them to achieve and maintain modest minds—minds "freed and cleansed," unencumbered by and dominant over the matter of their investigations. For Cavendish, in contrast, such instruments "intoxicate" rather than purify their minds. Furthermore, far from detaching philosophers from matter, microscopy fastens them to it while at the same time limiting them to the surface of things ("exterior figures of objects"), superficialities as opposed to deeper, internal, and thus valuable truths.

It is precisely such deeper truths that Cavendish seeks in her empirical practice, and she suggests that her own sensitive methods of observation and knowledge-making are more effective at accessing these truths than the more conventional methods employed and encouraged by Royal Society members like Hooke. Cavendish's critique of philosophers' reliance on optical technologies is part of a broader challenge she poses to contemporary empirical methods that had initially been inspired by Francis Bacon in his *Novum Organum* (1620). It is not the act of looking closely through magnifying glasses in and of itself that troubles Cavendish. Rather, it is what this act represents—namely, philosophers' delusional faith in their ability to see rightly at a distance and in what Keller portrays as their "perceptual passivity."[85] According to Bacon, the new science promised to allow men to "recover" their "right over nature." However, to achieve scientific authority and, thus, to fulfill this promise, Bacon insisted that men needed to perform an "expurgation of the intellect."[86]

Men "would be sufficient of themselves," Bacon observes, "if the human intellect were even, and like a fair sheet of paper with no writing on it. But since the minds of men are strangely possessed and beset, so that there is no true and even surface left to reflect the genuine rays of things, it is necessary to seek a remedy for this also."[87] Men's minds are plagued by "idols, or phantoms" that threaten to prevent them from perceiving clearly and accurately.[88] However, by teaching men how to cleanse their minds of these blots, Bacon seeks to "perform the office of a true priest of sense."[89] The mind must, according to Bacon, be "guided at every step; and the business . . . done as if by machinery."[90] Consequently, if the new science is going to produce the reliable knowledge that Bacon seeks, then philosophers must, according to Bacon, develop a means by which to resist what he portrays as the inherent tendency of humans to "mingle" with matter.[91] Their minds had to be "free and cleansed"—detached from as well as transcendent above the material world.[92]

According to Cavendish, Baconian modesty is a false presumption that encumbers scientific progress because it distracts philosophers from worthier methods that promise to produce real discoveries. "[A]s boys that play with watery bubbles or fling dust into each other's eyes, or make a hobbyhorse of snow, are worthy of reproof rather than praise, for wasting their time with useless sports," Cavendish explains, "so those that addict themselves to unprofitable arts, spend more time than they reap benefit thereby": "Nay, could they benefit men either in husbandry, architecture, or the like necessary and profitable employments; yet before the vulgar sort would learn to understand them, the world would want bread to eat, and houses to dwell in, as also clothes to keep them from the inconveniences of the inconstant weather."[93] For Cavendish, Baconian modest witnessing causes philosophers to waste their time; addicted to their own childish amusements, they lose touch with what really matters ("necessary and profitable employments"). This sentiment flies directly in the face of the enthusiastic assertions of public utility for which the early Royal Society was known. Take, for instance, Thomas Sprat's claim in his foundational *History of the Royal Society* that, as a result of its fellows' modest pursuits, England "has been able in a short time so well to recover it self: as not onely to attain to the perfection of its *former* Civility, and Learning, but also to set on foot, a *new* way of improvement of Arts, as *Great* and as *Beneficial* (to say no more) as any the wittiest or the happiest Age has ever invented."[94] Cavendish directly critiques such practical benefits of the new science, indicating that if true utility is to be had, then empirical philosophers must find methods for making knowledge other than those that Bacon sets out. She thus simultaneously challenges male arrogance and conventional Baconian practices of modest witnessing.

To delve further into Cavendish's *Observations*, with an eye to its expression of key vitalist materialist principles, is to discover why it is that Cavendish finds the practice of Baconian modesty so ineffectual. I would suggest that, for Cavendish, such modesty is based on a lie and is, therefore, infeasible. The modest "man conceits himself to be above nature," Cavendish observes, when in truth "all creatures of nature are produced but out of one matter, which is common to all, and . . . there are continual and perpetual generations and productions in nature, as well as there are perpetual dissolutions."[95] As Liza Blake has shown, in this passage and others like it from *Observations* that evoke what might be seen as a "commonality of matter" or *common nature*, Cavendish expresses a strong allegiance to vitalist materialism.[96] Evoking Epicurean transcorporeality in a vein similar to Hutchinson, her assertion that "flowing atoms enter and issue" via "pores" in all material beings directly corresponds to Lucretius's Epicurean portrayal of all beings as "confluence[s] of seeds" or atoms.[97] For Cavendish as for Hutchinson and Lucretius, all bodies are composites, or in the words of Cavendish "Composed Figures."[98] Cavendish also shares Lucretius's Epicurean conviction that,

motions doe
Dissolve all creatures, and their frames renew;
Whither first bodies, by impulsive force,
Or moving faculties, maintaine their course,
Wandring for ever in th'unbounded space.[99]

For Cavendish, as for Hutchinson and Lucretius, atoms are in a constant state of transcorporeal movement, and this movement continually renews the "frames" of "all creatures": "Creatures must be produced by Creatures. . . . Wherefore, all Natural Creatures are produced by the consent and agreement of many Self-moving Parts, or Corporeal Motions, which work to a particular Design, as to associate into particular kinds and sorts of Creatures."[100] Comprised not only of atoms but also of "vacuitie" or void, all bodies are porous and, with the perpetual movement of atoms, are therefore perpetually penetrated by the atoms of other bodies. Thus, there can be no autonomous being in the Epicurean universe: "no fixt seate can be for bodies found."[101] All beings are inseparable from their environments and, as Aaron Hanlon observes, "humans and nature [are] of the same cloth."[102]

The philosopher is no exception to this tenet; consequently, for Cavendish, Baconian modesty is based on a delusion if not an outright lie. According to Cavendish, not even the philosopher's intelligence or soul fundamentally distinguishes him from nonhuman nature: "there is [even] a soul and intelligence in the loadstone . . . as that there is a soul in man."[103] In other words, humans are of nature, not above nature, and only a "fool" believes otherwise.[104] Thus, Cavendish implies, it is foolish (and even deceptive) to cultivate the masculine modesty that Bacon recommends as a means of ensuring the authority of one's empirical practice, for it is predicated on detaching oneself from, and thereby transcending, matter, which is an impossibility. None of us—not even the modest virtuoso—has "power at all over natural causes and effects": "neither can natural causes nor effects be overpowered by man so, as if man was a degree above nature, but they must be as nature is pleased to order them; for man is but a small part, and his powers are but particular actions of nature, and therefore he cannot have a supreme and absolute power."[105] Empirical philosophers like Hooke and Henry Power, another microscopist and early member of the Royal Society Cavendish repeatedly targets, believe their modesty allows them to gain special access to nature: detached from and dominant over her, they are capable of, to quote Bacon, "entering and penetrating into [her] holes and corners" or subjecting her to "constraint" and "vexation," all in the service of fruitful experimentation.[106] In accordance with her persistent espousal of vitalist materialist principles, Cavendish disputes not only the practical benefits of this empirical approach but also its ontological foundation. For Cavendish, even the most virtuosic philosopher is fundamentally a material being and is therefore fundamentally incapable of "expurgating" his mind of matter as Bacon advises him to do.[107] As Keller explains, Cavendish insists "that man is inextricably *a part* of the nature he seeks to know: there simply exists no outside vantage point from which to view and thereby to control some object called nature."[108] Thus, whereas Hutchinson implicitly eschews the Baconian method of modest witnessing, Cavendish does so overtly. However, the challenges that both women present to Baconian empiricism are similar, largely because of their similar affinity for vitalist materialism.

The philosophical methods they propose as alternatives to Baconian modesty have striking similarities as well: they both use their material nature to their philosophical advantage. Rather than distancing herself from matter as Bacon encourages, Cavendish not only acknowledges her material nature but cultivates an affective openness to her objects of investigation, which in effect makes an asset out of her femininity. She summarizes the method she proposes thus: "the best optic is a perfect natural eye, and a regular sensitive perception; and the best judge, is reason; and the best study is rational contemplation joined with the observations of regular sense."[109] Cavendish advocates a bipartite philosophy, comprised of both "sensitive" and "rational" perception. "A regular sensitive perception" provides the foundation, while reason is the arbiter. Though, as Lisa Sarasohn observes, Cavendish emphasizes "perception['s] inability to comprehend the 'interior, proper, and innate actions' of a creature," rather than eschewing sense perception and the physical engagement with matter that it entails, she insists on the value and importance of engaged affective observation, in spite of its limitations.[110] Whereas Bacon promotes an empiricism that is (or at least purports to be) pure of material entanglement and therefore aloof and sober, Cavendish's empiricism is thoroughly enmeshed in matter and therefore receptive and vulnerable to stimuli. Though they both share an investment in the importance and value of experience to empirical philosophy, while the Baconian witness strives to be modest to transcend the idols of the mind, Cavendish's witness strives to be *sensitive*.

Take, for instance, her "Observation X. Of a Butterfly." Its opening is indicative of its approach: "Concerning the generation of butterflies, whether they be produced by the way of eggs, as some experimental philosophers do relate, or any other ways; or whether they be all produced after one and the same manner, shall not be my task now to determine; but I will only give my readers a short account of what I myself have observed."[111] This is far from a deductive or speculative method. Indeed, if anyone is speculative in their reasoning in this passage, it is the "experimental philosophers" whose theories seem to Cavendish unsubstantiated—based purely on what has been related—in comparison to hers, which she portrays as firmly rooted in personal ("I myself") experience. Though she does not demand replication for corroboration and validation like Bacon does, implying that her sensitive witnessing is in and of itself enough to verify truths, Cavendish's method is distinguished by induction much as Bacon's is.[112] This becomes all the more apparent as she proceeds to provide her account, which begins thus: "When I lived beyond the seas in banishment with my noble lord, one of my maids brought upon an old piece of wood, or stone (which it was I cannot perfectly remember) something to me which seemed to grow out of that same piece; it was about the length of half an inch or less, the tail was short and square, and seemed to grow out of that same piece."[113] She proceeds to amass multiple physical details about this and another similar specimen, which she explains, "I laid by upon the window; and one morning I spied two butterflies playing about it; which, knowing the window had been close shut all the while, and finding the insect all empty, and only like a bare shell or skin, I supposed had been bred out of it."[114] As these passages illustrate, Cavendish relies as heavily on "observations of regular sense" as Baconian empiricists do.

In fact, her refusal to employ a microscope makes such unaccoutred observations all the more important. Whereas Robert Hooke or Henry Power have recourse to the microscope, Cavendish instead trains her "perfect natural eye" sensitively on the objects of her investigation, readily acknowledging the limit of her sight while nevertheless celebrating the reach of her observation: "But yet this latter I will not certainly affirm, for I could not discern them with my eyes, except I had had some microscope, but a thousand to one I might have been also deceived by it: and had I opened this insect, or shell, at first; it might perhaps have given those butterflies an untimely death, or rather hindered their production."[115] Cavendish here sets clear limits for her empirical practice, eschewing both microscopic investigation as well as dissection that would result in the specimen's death. She then proceeds to decline either to trust or to challenge empirical hearsay: "I have heard also that caterpillars are transformed into butterflies; whether it be true or not, I will not dispute."[116] Instead, she concludes by returning to her own sense perception, conveying an attentive observation of caterpillars' behaviors: "only this I dare say, that I have seen caterpillars spin, as silkworms do, an oval ball about their seed, or rather about themselves."[117] We might be tempted to presume that the emphasis Cavendish places on herself and her own observations in such passages implies a firm if not fixed subjectivity that contradicts the openness and vulnerability that her practice of sensitive witnessing entails. However, analyzed through the framework of her sustained vitalist materialism, her repeated use of the "I" takes on a different valence. It marks her as merely a creature among creatures whose physical and psychological boundaries are provisional at best. Hers is a humble rather than assertive subjectivity, which is evinced by how she prefaces her observations—"only this I dare say"—and this humility extends to the sensitive perception she practices and promotes. She is engaged with and responsive to rather than detached from the specimens she observes. Indeed, whereas such engagement may be perceived to undermine the modest witness's authority, within the context of her unorthodox experimental philosophy, it bolsters her authority, attesting to the precision and veracity of her observation.

As Cavendish's empirical observations convey, though she may at points emphasize the crucial role she believes reason to play in philosophy, "sensitive perception" nevertheless retains a prominent position within her philosophic practice. Throughout her writings, Cavendish demands that we resist the temptation to polarize matter and mind, sense and reason, insisting on their inseparability. While she may enjoy imagining a "Blazing World" that is "composed only of the rational" parts of matter, we betray the nuances of her philosophical method (not to mention the creative liberty asserted by her fiction) if we impose her fantasy of the fictional Duchess's world-making on her *Observations upon Experimental Philosophy*.[118] These works were originally published in the same volume, but their differences are at least as instructive as their continuities are. In fact, in an essay entitled "Of the Senses and Brain" included in *The Worlds Olio*, she directly refutes the plausibility of the fiction she presents in *The Blazing World*:

> Some say, That there is such a nature in Man, that he would conceive and understand without the *Senses*, though not so clearly, if he had but Life, which is Motion. Others say, There is nothing in the *Understanding*, that is not first in the *Senses:* which is more probable: for the *Senses* bring all the Materials into the *Brain*; and then the *Brain* cuts and divides them, and gives them quite other Forms than the *Senses* many times have presented.[119]

Here again, we see Cavendish presenting a bipartite method of observation. In reality, according to Cavendish, sense and reason are dialectically intertwined and, thus, fundamentally resistant to any attempts to polarize them. As she goes on to observe in *The Worlds Olio*, "we cannot call that a perfect knowledge which our *Reason* singly tells us; but what our perfect and healthful *Senses*, joined with our *Reason*, distinguish to us."[120] So intimate is their collaboration that, in Cavendish's philosophical practice, sense and reason do not only operate in conjunction with one another but mutually inform and even help constitute one another.

We have thus far confined our attention to the sensitivity with which Cavendish engages in physical observation. However, I would contend that her rational explorations are sensitive as well, albeit differently so. In *Observations*, she celebrates the superiority of rational "discourse" to "deluding arts" (such as microscopy), attributing to the former the ability to "find or trace nature's corporeal figurative motions" while accusing the latter of being incapable of sufficiently "inform[ing] the senses."[121] With the word "trace," Cavendish imbues reason with physicality, even making it seem like a haptic activity: to employ reason well is not simply to manufacture ideas out of thin air but, rather, to follow the motions of nature as if by foot or by hand.[122] In other words, reason must, like sense, be sensitive to nature for it to fulfill its promise of providing insight. Cavendish's impulse to physicalize reason and to promote a cultivation of its sensitivity extends to *The Worlds Olio* as well. There, she likens reason to the eyes: "for, the *Brains* are like Eyes, of which some are so quick, that they cannot fasten upon an Object to view the perfection of it. Others so dull, that they cannot see clearly; or so slow, they cannot untie themselves soon enough, but dwell too long upon it. So it is the discussing of the Object well, that encreases or begets Knowledg."[123] We must discuss and reason about "the Object well" if we are going to make knowledge effectively. To do so, we must aspire to a reason that is neither so quick that it is unable to "fasten upon an Object" nor so slow that it is unable to "untie" itself from the object. The proposal here is for a methodical and yet softened reason: a reason that, operating in a relaxed and open fashion, is highly receptive and responsive to the objects of its explorations. Here, as in *Observations*, Cavendish celebrates and advocates a sensitive reason that closely corresponds to and even echoes the sensitive perception she also recommends—a reason that is "acutely affected by external stimuli."[124] Hers is a distinctively feminine reason distinguished by softness, tenderness, and delicacy.

The Sublime Sensitivity of Cavendish and Hutchinson

The "acuteness" of the affective experience around which both Hutchinson and Cavendish organize their philosophical methods and the knowledge-making these methods produce associate them both with the early modern sublime. On the one hand, the experiences both philosophers depict are widely variable—as variable as the objects they investigate and explore. To be affectively stimulated by "prodigious thunders," like those Hutchinson portrays in her depiction of the deluge, is certainly distinct from being affectively stimulated by the sight, depicted by Cavendish, of "caterpillars spin[ning], as silkworms do, an oval ball about their seed."[125] Nevertheless, I would propose that, in fact, the affective experience entailed by the sensitive witnessing Hutchinson and Cavendish envision is not as variable as it might superficially seem. The consistency of their methods produces a consistency of affect. Indeed, underlying all their observations, explorations, and investigations is a persistent affective experience of sublimity. Given the attention that the epic poem devotes to devastating Biblical events, such sublimity is readily discerned in Hutchinson's *Order and Disorder.* However, it recurs throughout Cavendish's philosophical works as well, albeit less explicitly. Take, for instance, this passage from "Fire compared to Stings," a verse Cavendish includes in *Poems, and Fancies*:

> Nothing is so like Fire, as a Flies Sting,
> If we compare th' effect which both do bring.
> Were there as many Stings, as Fiery Atomes small,
> Would pierce into the Flesh, Bones turne to Ashes all.
> Thus we finde Flies do carry every where
> Fire in their Tailes, their Breech they do not feare.[126]

Here, as Cavendish likens the sting of a fly to fire, she portrays it as intensely affecting, despite its minuteness. To know the fly's sting fully, the philosopher-speaker experiences it as sublime, even envisioning it turning her bones into ashes.

Cavendish's poem, "A World in an Eare-Ring," similarly illustrates the association between her practice of sensitive witnessing and the sublime while also emphasizing its association with femininity. The poem begins thus:

> An Eare-ring round may well a Zodiacke bee,
> Where in a Sun goeth round, and we not see.
> And Plantes seven about that Sun may move,
> And Hee stand still, as some wise men would prove.
> And fixed Stars, like twinkling Diamonds, plac'd
> About this Eare-ring, which a World is vast.[127]

Depicting a minute plurality of worlds embodied by a decorative feminine bauble, this poem expresses Cavendish's investment in Epicurean atomism while also emphasizing the compatibility between the practice of sensitive witnessing she advocates and femininity. Cavendish suggests that, conventionally associated with trivialities (in fashion and in domesticity), women are particularly well situated to perceive, through their sensitive reason (if not their sensitive perception), the minute sublimity ("a World . . . vast") that such sensitivity uncovers. As the poem goes on to observe, "And Lightnings, Thunder, and great Winds may blow / Within this Eare-ring, yet the Eare not know," it emphasizes the sublimity available to the witness who employs her sensitive reason. Though the "Lightnings, Thunder, and great Winds" that "blow" through the bauble are imperceptible to "the Eare," the speaker is nevertheless capable of witnessing them. I would argue that it is the speaker's sensitive reason that allows her to extend her perception beyond that of her ear and, thus, to perceive such minute sublimity. The poem suggests that such sublimity is likely to go unremarked by the modest virtuoso who is determined to maintain both a modest distance from matter as well as his masculinity. Such a virtuoso is, Cavendish suggests, likely to remain oblivious to such evasive minute worlds. In contrast, a female philosopher like herself, who is closely associated with and sensitive to material trivialities, is particularly well poised to detect them.[128] Cavendish thus transforms femininity into a superpower, capable of allowing the philosopher to boldly go where men are disinclined, if not unequipped, to go.

Admittedly, it is anachronistic to apply the term *sublime* to the sensitive witnessing promoted by Hutchinson and Cavendish. As Jonathan Lamb observes, though Boileau's French translation of Longinus's first-century treatise *Peri Hupsous* in 1674 set an important foundation for the modern popularity of the sublime, "it was not until [Leonard] Welsted's in 1712 (reprinted in 1724) and [William] Smith's popular *Dionysius Longinus on the Sublime* (1739) that Longinus and the sublime became thoroughly current in Britain."[129] It would take even longer for the sublime to expand beyond its initial status as a rhetorical effect, which it predominantly is in *Peri Hupsous*, to include effects provoked by intense lived experiences (e.g., awe-inspiring encounters with vast natural landscapes). The knowledge-making promoted by Hutchinson and Cavendish intimates both the rhetorical and the experiential versions, despite their historical antecedence to the modern flourishing of the aesthetic category of the sublime. Boileau says of Longinus that, "Often he uses the very figure he is discussing, and in talking of the sublime, is most sublime himself."[130] The same is true of Hutchinson and Cavendish. They regularly achieve poetic sublimity as they engage in describing episodes and characteristics, be they epic or trivial. The deluge that Hutchinson describes is sublime, but so too is her poetic depiction of it. The same is true of Cavendish's poetic depiction of the disintegration of human bones by fire and a world in an earring. Therefore, their practices of sensitive witnessing are appropriate starting points for this historical study which seeks to excavate the persistent relationship between sensitive witnessing and sublime knowledge-making across the long

Enlightenment period. They illustrate the close connection between sensitive witnessing and femininity, which enabled transformation of the latter from a liability into an asset.

• • •

Historians and cultural theorists regularly attribute man's alienation from and dominance over nonhuman nature to the Enlightenment era.[131] Bacon certainly aspired to a philosophic mind that had been purified of its affiliation with matter. Such purification was an important practice for the modest witness whose authority was contingent on his ability to deny his own intermingling with matter. Informed by the Epicurean materialism poeticized by Lucretius, Lucy Hutchinson and Margaret Cavendish envision a distinctive alternative—the sensitive witness. Careful observation was, of course, a key component of Baconian empiricism. However, though early Baconian philosophers not only carefully observed but also engaged in physical manipulation of the objects of their investigation, they simultaneously downplayed the immersive materiality that such experimentation entailed and the material continuity between man and matter that it registered. To preserve their modesty, they felt compelled to counteract the materiality of their empirical endeavors with a determined assertion of their own separation from matter. Hutchinson and Cavendish sought a different path to philosophic authority. Rather than asserting their separation from matter, they emphasized their connection with it and, in developing and promoting a method of sensitive witnessing, seized the philosophical advantages this connection afforded.

Bacon manifested an interest in Epicurean materialism but, over the course of his career, grew increasingly ambivalent and, ultimately, doubtful about its legitimacy.[132] Hutchinson's and Cavendish's attraction to Epicureanism was more persistent; they both continued to be influenced by Lucretius even after they had begun to distance themselves from certain Epicurean principles. The transcorporeality and material continuity between humans and nonhuman nature that Epicureanism recognized helped establish the foundation for the sensitive witnessing that they pioneered. Epicureanism inspired the sensitive witness's practices and reinforced her knowledge-making potential. While Baconian modesty was not easily obtainable for the female philosopher because she was perceived to be too closely associated with matter to be able to achieve it, sensitivity, on the other hand, was readily available to her precisely because of her perceived intimate connection to matter. Thus, with the method of sensitive witnessing, Hutchinson and Cavendish did not simply develop an innovative philosophical method but developed one that was, unlike Bacon's more orthodox variety, accessible to women as well as men.

According to Bacon's portrayal of the modest witness, her close association with the body and, thus, with matter would put the female philosopher at a significant disadvantage, even potentially precluding her from engaging in respectable philosophical practice altogether. However, Hutchinson and Cavendish instead chose to view such material association through the framework of vitalist materialism. As a result, their affinity with matter becomes something altogether different. Epicurean philosophy normalizes the irresistible pull of one's material environment and insists that all beings (whether female or male) are equally exposed to it. Women are simply more accustomed to acknowledging this pull because society demands that they do so. I agree with Keller's observation that Cavendish "did not believe that *as a woman* she had access to any privileged epistemological standpoint."[133] However, I would suggest that, because of their recognition of how society constructs gender and associates women with matter, both Cavendish and Hutchinson did view themselves to be better positioned and equipped as women to reap the great benefits of sensitive witnessing than male philosophers were. It is a privilege to deny one's material nature—a privilege to which only free men had access. Rather than defying gender expectations and attempting to claim this privilege for themselves, Hutchinson and Cavendish chose instead to reveal such privilege to be unwarranted and founded on a deceptive fiction about the status of man in relation to matter. Their vitalist materialism substantiates their insistence that men are no different than any other creature—they too are porous composites, entangled with their environments through transcorporeal atomic exchange. Hutchinson and Cavendish pursue and promote sensitive rather than modest witnessing not as a compromise with patriarchal expectations and capitulation to the female gender role but, rather, because sensitivity is, for both female philosophers, more compatible with, and thus equipped to access, the cosmos than modesty is. To return to this chapter's epigraph, for both Hutchinson and Cavendish it is the empiricist who cultivates rather than eschews softness—openness, vulnerability, sensitivity—that has the potential to restore a "rusty" age and is therefore well prepared to enact a truly "great instauration."

TWO

LIBERTINE EPICUREANISM AND SCIENTIFIC CRITIQUE: APHRA BEHN

THE SATIRE OF SCIENCE that Aphra Behn's *The Emperor of the Moon* presents is as biting as it is hilarious, yet the precise nature and results of its satire are not as straightforward as one might expect. A farce composed in the commedia dell'arte tradition, *Emperor* was "adapted to our English Theatre and genius" from scenes included in *Arlequin empereur dans la lune*, a performance first staged in Paris in 1684.[1] In her portrayal of the character Doctor Baliardo, Behn invites her audience to laugh at a virtuoso whose philosophical enthusiasm has gotten the better of him. As early as scene 2, the Doctor is readily convinced that he has been able to spy on the emperor of the moon through his "telescope twenty (or more) foot long."[2] The farce culminates with a descent of the figures from the moon-world to earth, strategically fabricated by the Doctor's daughter Elaria, niece Bellemante, and their lovers Don Cinthio and Don Charmante to trick him into allowing their marriages. The Doctor's excessive credulity is clearly a main target of Behn's comedic satire. However, what exactly lies behind, and is signified by, his credulity requires deeper investigation.

Like Hutchinson and Cavendish, Behn drew significant inspiration from Lucretius's *De Rerum Natura*. By the time she composed *The Emperor of the Moon*, most likely in 1684, "the neo-Epicurean revival" was well under way.[3] Thomas Creech had published his acclaimed translation of *De Rerum Natura* in 1682, and Behn was such an admirer of it that she would publish the laudatory poem "To Mr. Creech (under the Name of *Daphnis*) on his Excellent Translation of *Lucretius*" in her 1684 *Poems upon Several Occasions*.[4] As Alvin Snider observes, "Creech's translation, which went through five editions by 1700 and made the Roman poet canonical among English readers in his generation, helped to put into general circulation the metaphysical view that to this day we associate with modernity: the belief that behind the illusion of a stable human nature lies only atoms and desires, all floating in the void."[5] Snider and other scholars have demonstrated that traces of such Epicurean beliefs may be found throughout several of Behn's works. Snider attends specifically to the Epicurean elements of Behn's poem "To Mr. Creech."[6] Mary Trull portrays Behn's roman à clef *Love-Letters Between a Nobleman and His Sister* (1684–1687) as a response to Lucretius, and Sarah-Gray Lesley traces the Roman poet's influence throughout Behn's novella *Oroonoko*.[7] Ros Ballaster and Robert Markley take a broader view. Ballaster analyzes a number of Behn's poems as well as her play *The Rover* in relation to Lucretian Epicureanism.[8] Markley refers to Lucretius as he contextualizes the radical libertine skepticism regarding both science and Christianity that he argues Behn expresses throughout several of her works (poems, as well as *Oroonoko* and her translation of Fontenelle's *Entretiens sur la pluralité des mondes*). Though Epicureanism is not a primary focus of Markley's essay, he nevertheless emphasizes the importance of the "kind of praise of Lucretius" her poem "To Mr. Creech" expresses, characterizing it as "a hallmark of theological skepticism and philosophical radicalism throughout the late seventeenth and early eighteenth centuries."[9] According to Markley, Behn's poem to Creech offers one of several various "instance[s] of her critique of both the beliefs and methodology that inform seventeenth-century efforts to theologize the natural world."[10] This chapter will further develop our understanding of how Behn's satire of contemporary philosophical practice coincides with her adherence to Epicurean principles. Markley concludes his analysis with the claim that Behn's "radical skepticism" is "defined by [her] ironic self-distancing."[11] Building on Markley's analysis, I argue that Behn not only critiques contemporary philosophical practice but also proposes an alternative practice that is informed by and thus compatible with her libertine Epicureanism. Joining Hutchinson and Cavendish, albeit with her own unique motivations and results, Behn's Epicureanism leads her to promote a practice of sensitive witnessing that she insists is far more feasible and reliable than the modest witnessing sponsored by the Royal Society.

Charmed by Lucretius

Her poem to Creech "on his Excellent Translation of *Lucretius*" shows that Behn's gratitude for his translation of *De Rerum Natura* is multifaceted. She begins by expressing her admiration for both "[w]orthy *Divine Lucretius*, and *Diviner*" Creech as poets.[12] However, it is not purely for their "Flights of Verse" that Behn appreciates them. She also celebrates, and implicitly endorses, the Epicurean philosophy that Creech has made available to her through his translation:

> The Mystick Terms of Rough *Philosophy*
> Thou dost so Plain and Easily Express,
> Yet Deck'st ‘em in so soft and Gay a Dress,
> So Intelligent to each Capacity;
> That They at once Instruct, and charm the Sense
> With heights of Fancy, Heights of Eloquence;
> And Reason over all unfetter'd Plays,
> Wanton and undisturb'd as Summers Breeze
> That gliding Murmurs o're the Trees,
> And no hard notion meets, or stops its way;
> It Peirces, Conquers, and Compells
> As strong as Faiths resistless Oracles,
> Faith the Religious Souls content,
> Faith the secure Retreat of Routed Argument.[13]

As these lines illustrate, Behn finds much to love in Lucretius's *De Rerum Natura*. Its "Mystick Terms" are instructive as well as charming. It stimulates not only her aesthetic appreciation and imagination ("Fancy") but also her intellect ("Reason"). Indeed, these lines indicate that her encounter with *De Rerum Natura* has strengthened her reason to such an extent that it has become at least as robust as the "faith" of "the religious." Behn therefore appears here to imply that she has been persuaded by the logic of the Epicureanism that Lucretius presents. In fact, Behn seems even more favorable to Lucretius than Creech does. In his introduction to his translation, Creech takes great pains to distance himself from Lucretius's Epicurean atheism, suggesting that one of his primary motives in translating *De Rerum Natura* is "to expose" and thereby shatter "the principles of irreligion" that the poem conveys.[14] Behn feels no such compulsion. On the contrary, she writes as if she has no reservations whatsoever about the plausibility and even persuasiveness of the materialist philosophy that *De Rerum Natura* embodies.

Mary Trull affirms Behn's endorsement of Epicurean atomic theory, demonstrating how she adopts Lucretius's view of the chaotic and seductive nature of an atomic universe in several of her works. Trull argues that *Love-Letters Between a Nobleman and His Sister* presents a particularly suggestive engagement with Epicureanism. According to Trull, "[t]o the reader of Creech's Lucretius, love is the most arbitrary force of all in a universe built upon atomistic anarchy; such is the vision of love that Behn offers in Part Three of *Love-Letters*."[15] Trull's analysis shows that Behn's acceptance of Epicurean principles runs deep: though she challenges the misogyny she detects in *De Rerum Natura*, she nevertheless adapts Lucretius's "analysis of agency and desire" to her purposes while "reflecting a deeply Epicurean view of the absolute freedom—even arbitrariness—of the will."[16] Influenced by Creech, Behn is, as Trull demonstrates, particularly influenced by the Epicurean notion of our "susceptibility" to impressions.[17] For Behn, "the general state of the . . . soul is one of vulnerability to constant incursions from without."[18] Thus, like Lucy Hutchinson and Margaret Cavendish, Behn emphasizes that no one is immune to atomic penetrations. However, according to Trull, as she follows Creech's Lucretius in insisting on the seductive quality of the "forms" or "images" that atomically produce sensation, Behn becomes particularly interested in the potential for atomic penetration to produce delusion.[19] Drawing on the quote below, Trull argues that, in Creech's translation of *De Rerum Natura*, "[S]eeing is loving" and "Lucretian images are by nature deceitful and coercive":

> But now from *Beauty*, now from *Forms* that please,
> What comes but *thin* and *empty Images*?
> E'en such as *He* enjoys that drinks in dreams,
> His *Thirst* encreaseth 'midst the *fancy'd* streams,
> So *Love* deludes *poor men*, their *covetous* Eye,
> What *long*, what *frequent* Sights can satisfye?[20]

Ros Ballaster similarly highlights how Behn's adherence to Epicurean principles inspires her to associate sense perception with the arts of seduction. As Ballaster shows, Behn regularly portrays sensory stimulation—whether effected by poetry or material form—as that which "charms" the observer. " 'Charm' is a term," according to Ballaster, "that chimes repeatedly in Behn's amatory verse":

> It is a term that bridges the philosophical, the physical and the poetic: it is associated with number, with music and with the involuntary or mechanical power of attraction through "form" itself. The "charms" of physical form and the charms of poetic form call each other up in Behn's poetry repeatedly, and poetry itself becomes a charm—in the sense of a verse to chant or recite with special powers of influence and attraction.[21]

As both Ballaster and Trull demonstrate, according to Behn's Lucretian Epicureanism, we experience perception when we are seduced or "charmed" by what we sense. We are therefore vulnerable to being moved, and potentially beguiled, against our will by sensory stimuli.

The Emperor of the Moon

Behn places such vulnerability at center stage in her farce *The Emperor of the Moon*. In his article "Retraining the Virtuoso's Gaze: Behn's *The Emperor of the Moon*, the Royal Society, and the Spectacles of Science and Politics," Al Coppola makes a persuasive argument that *Emperor* is, "above all, an interrogation of improper spectating" and that, with the farce, Behn "is seeking to teach a rational mindfulness to a society overrun with enthusiastic virtuosi."[22] Contending that "[i]t was the endemic abuse of spectacle for political ends in the 1680s that Behn's play sought to curb," Coppola associates *The Emperor of the Moon* with "the efforts of the Royal Society in its own printed works of the 1680s—most notably [Nehemiah Grew's] *Musaeum Regalis Societatis*—to retrain the virtuoso's gaze and inculcate a crypto-Tory political agenda."[23] According to Coppola, the *Musaeum* "works to produce a reader inoculated against enthusiasm" and so too does Behn's *Emperor*.[24]

While Coppola goes a long way in illuminating what common ideologies and political concerns motivate both the Royal Society and Behn to take aim at philosophical enthusiasm, I would like to build on his analysis by attending to what distinguishes Behn's *Emperor* from Grew's *Musaeum*, for in doing so we can begin to appreciate the uniqueness of Behn's philosophical approach. Both works may strive to restrain virtuosic spectating, but their techniques for achieving such restraint vary significantly. Indeed, I would argue that, though they may be informed by similar ideological motivations, Behn would not have been satisfied by the solution to "improper spectating" Grew and his Royal Society colleagues proposed. *Emperor* presents a serious challenge to the efficacy of the solution to the problem of philosophical enthusiasm that the *Musaeum* offers.

Behn may, as Coppola suggests, share with Grew a desire to "inoculate against enthusiasm"; however, *The Emperor of the Moon* not only neglects to endorse the precise method of inoculation that Grew pursues but, in fact, lampoons it. To appreciate this dimension of her satire, we must recognize the various ways in which Doctor Baliardo himself performs natural philosophy in a manner that evokes the very philosophical methods Grew represents. We first meet the Doctor in act 1, scene 2 as he enters "with all manner of mathematical instruments hanging at his girdle; [the Doctor's servant] Scaramouch bearing a telescope twenty (or more) foot long."[25] Set in the context of a farce, this entrance immediately invites laughter with the excessive number of the various "mathematical instruments" and the excessive length of the telescope affording comedy through exaggeration. However, placed alongside Grew's *Musaeum*, Behn's stage directions seem more realistic than hyperbolic. Grew devotes an entire section of the *Musaeum* to "Things relating to the MATHEMATICKS; and some MECHANICKS." This section begins with a discussion of "a REFLECTING TELESCOPE. Contrived by Mr. *Isaac Newton*, Professor of the *Mathematicks* at *Cambridge*" and then proceeds to catalog various other instruments, including "An Instrument for working Questions by Multiplying and Dividing," "A WAY-WISER," several architectural models, "An INSTRUMENT, contrived by Sir *Christopher Wren*, to demonstrate, How far against the Wind a Ship may Sail," "A TERELLA, or an Orbicular *Loadstone*," and "Two DIPPING-NEEDLES."[26] Grew might as well be wearing these miscellaneous instruments "hanging at his girdle" just like Doctor Baliardo does. When such similarities with this section of Grew's catalog are recognized, Behn's satirical introduction of the Doctor appears to parody, and thereby criticize, rather than affirm the *Musaeum* and the philosophical practice it represents. That Grew expresses a wish, in the *Musaeum*'s dedicatory epistle addressed to Royal Society secretary Daniel Colwall, that members of the society "might always wear this Catalogue, as the Miniature of your abundant Respects, near their Hearts" only further substantiates *Emperor*'s satirical relationship to the *Musaeum*.[27]

Other aspects of Behn's satirical portrayal of Doctor Baliardo also closely correspond to Grew's *Musaeum*. In one of the Doctor's first lines, which he utters to his servant Scaramouch while discussing the power of his telescope to "let the sight into the secret closet" of "the great monarch of the upper world," he claims, "I take survey of all."[28] It may be the case, as Coppola's essay suggests, that Grew would join Behn in criticizing the Doctor's particular use of his telescope as a "statesman's peeping-hole" and thus as a tool for gaining access to the monarch's affairs. However, Baliardo's claim for the encompassing nature of his observation—his ability to examine "*all*"—is matched rather than denounced by Grew's *Musaeum*. Perhaps the most obvious feature of his catalog, and the Royal Society's collection that serves as its basis, is its comprehensiveness. With their collection and catalog, Grew and his fellow society members seem, like Baliardo, to be capable of "surveying all." Just a glance at the *Musaeum*'s table of contents ("Prospect of the whole WORK of the *Musaeum*") gives this impression. It is divided into four parts—"Of Animals," "Of Plants," "Of Minerals," "Of Artificial Matters"—and each part is divided into multiple sections. The specimens it includes have been collected not only from England but also from various distant lands: a mummy from Egypt, a "GREAT SHELL'D HEDGHOG" from Brazil, the "HORNES of a WILD BULL" from Africa, a cow's tail from India, a "LANTHORN-FLY" from Peru, "COD of the wild LOCUST" from Virginia.[29] It is as if the Royal Society has provided the world inside a collection. Thus, as Grew depicts the collection in painstaking detail, he appears to afford his reader the opportunity to, like Behn's Doctor Baliardo, "survey all."

For our purposes, though, the most important feature that Grew's and Behn's portrayals of philosophical practice share is their emphasis on modest witnessing, with the former advocating it and the latter satirizing it. The *Musaeum* is a masterwork of Baconian modesty. As Coppola shows, even the "Skeleton of a crocodile or ye leviathan" is prevented from violating the philosopher's reserve: "Through Grew's anatomical method and the plate's arresting dislocations of scale, the awfulness of Leviathan, which has been so assiduously summoned for the reader, is nevertheless carefully contained: its bulk measured, its bones counted, all of it cut down to size."[30] Throughout the *Musaeum*, the philosopher is shown to hold its objects of investigation at arm's length, employing a range of distancing techniques to prevent himself from appearing to become entangled with them. Take, for instance, Grew's depiction of a "STONE voided from the *Penis* or *Urethra* of a Man who lived at *Exeter*. Given by Dr. *Cotton*":

> It is of a whitish colour, and soft substance, almost like Chalk. In length two Inches and a quarter. Of a Pyramidal figure; with an obtuse Cone. Near the Base an Inch over. Where it hath a little Hole or Canale tending towards the Cone. When it first slipped out of the Bladder into the *Penis*, it was neither so thick or big, nor so hard, but that, as it seems, the Urine

> pressing forward, forced a hole for its passage through the middle of it. Which being opened, the Stone continued fixed in the same place, *viz.* about an Inch behind the *Glans Penis*, for the space of Thirteen Years. In which time, it gradually grew bigger, till it came to the bulk above mention'd. And the said Hole or Canale being by the continual accretion of new matter, at last stop'd up, the Stone was then forced out of the end of the *Penis*.
>
> This Man, in all this time, scarce felt any great Pains; neither did he omit his usual Recreations or his business. And once he took a Journey (*on Horseback*) from *Exeter* to *London*, is [*sic*] about an hundred and thirty eight miles, without any trouble."[31]

I have chosen this particular passage and have quoted it at length to show just how effectively Grew manages to sustain his Baconian modesty. The story it tells of a man voiding a stone, over two inches in size, naturally through the end of his penis after having unknowingly (and without "any great Pains") retained it within his penis for thirteen years is arresting. Nevertheless, the author retains a neutral style throughout his depiction and narration which denies the palpability of the details that he provides. In spite of the visserality of the passage's content, the tone remains detached and the overall style aloof as they do throughout the rest of the book. The geometrical descriptors ("Pyramidal," "obtuse Cone") and measurements ("in length two Inches and a quarter," "about an Inch behind the *Glans Penis*") contribute to such detachment by transmuting the physicality of the specimen into abstraction. Grew's Baconian modesty appears to contain the specimen, restrain the philosopher's gaze, and thus maintain a separation between the philosopher and the object he investigates such that he is able to remain unsullied by matter.

Behn parodies precisely such philosophical modesty. As Baliardo engages in a discussion with Charmante, having taken him for a member of "the great cabala—of Eutopia," he inquires how he might be granted membership into the "cabala." Charmante advises him to preserve his purity and, thus, his philosophical modesty:

> CHARMANTE: Were you but once admitted into that society—
>
> DOCTOR: Aye, sir, what virtues or what merits can accomplish me for that great honour?
>
> CHARMANTE: An absolute abstinence from carnal thought, devout and pure of spirit; free from sin.

DOCTOR: I dare not boast my virtues, sir; is there no way to try my purity?[32]

Charmante proceeds to pretend to put the Doctor's "strength of virtue" to a trial. "If you be thoroughly purged from vice," Charmante explains, "the optics of your sight will be so illuminated, that glancing through this telescope, you may behold one of those lovely creatures, that people the vast region of the air."[33] During the trial Charmante advises the Doctor to "shake off all earthly thoughts" and rewards Baliardo for his success at doing so with "a picture of a nymph," which he places at the end of the telescope for the Doctor to take for a real view of the moon.[34] Having supposedly relinquished "carnal" or "earthly thoughts," Baliardo believes himself to have been able to "purify" and "purge" himself, becoming detached from and ultimately transcending the material world. Thus, he perceives himself to have succeeded, like Grew, at becoming a modest witness. It is no wonder that he believes he can view the people of the moon. As a modest witness, he is no longer of this earth. His explorations are free to soar where they will. Appearing to fully relinquish his tie to the material world, he seems truly capable of "surveying all."

Of course, in this scene as throughout the rest of the play, Behn invites her audience to laugh at Baliardo's philosophical practice, and I would argue that, as she does so, she satirizes the practice of modest witnessing in which works like Grew's *Musaeum* engage. The dialogue that follows the Doctor's encounter with the sight Charmante has presented to him emphasizes the limited value of philosophical modesty:

DOCTOR: Astonished, ravished with delight, I see a beauty young and angel-like, leaning upon a cloud.

CHARMANTE: Seems she on a bed? Then she's reposing, and you must not gaze.

DOCTOR: Now a cloud veils her from me.

CHARMANTE: She saw you peeping then, and drew the curtain of the air between.

DOCTOR: I am all rapture, sir, at this rare vision.[35]

As this passage demonstrates, Baliardo's modesty is incapable of moderating his vulnerability to wonder and desire. In spite of having just ostensibly "purified" himself and transcended materiality, he cannot help but be "enraptured" by what he sees. In fact, he is "ravished" by the image, a word choice that suggests that he is not simply moved but penetrated by it.[36] Thus, Behn indicates that Baliardo's encounter with the "angel-like" woman "leaning upon a cloud" has undermined not only his transcendence of matter but also the enclosure and, thus, integrity of his very being. So titillated, Baliardo loses himself as well as his composure. This progression from modesty to discomposure transpires again, in the same scene, when Charmante fools the Doctor into believing he has seen the moon's emperor. When Baliardo tells Charmante that he has not yet seen the emperor, "the mighty Iredonozar," Charmante directs him thus: "Refine your thoughts, sir, by a moment's prayer, and try again." As the Doctor follows his recommendation, Charmante "*claps the glass with the emperor on it* [*onto the telescope*]." Baliardo is fooled again into believing the counterfeit sight and responds, again, with heightened emotion: "It is too much, too much for mortal eyes!"[37] With this scene, Behn insists that modesty is no match for matter's charms.[38]

While it is true that Doctor Baliardo presents an extreme, and thus particularly humorous, case, to compare *Emperor* to Grew's *Musaeum* is to recognize that the confidence and even arrogance that provide the context for his credulity accord with what was viewed as acceptable, even admirable, philosophical practice during the Restoration period. Behn emphasizes the relevance of her portrayal of the Doctor to contemporary trends by making certain threats to the effectiveness of his philosophical practice—specifically, the tendency of sense perception to mislead and/or stimulate the observer—recurring themes throughout the play. It is not just the Doctor who is prone to the effects of these threats. While he may experience them the most dramatically, other characters are also confronted with the difficulty of attaining reliable information from the senses and with the inclination to be titillated by what they perceive. For instance, at the beginning of the second act, a confusion in identity occurs when, to avoid her father's detection, Elaria hides her lover Cinthio in Bellemante's closet. In the scene before, unbeknownst to Elaria, Scaramouch had hidden Bellamante's lover Charmante in the same closet to avoid Baliardo's detection, and when Cinthio is surprised to encounter the form of Charmante in the dark, he mistakes his friend for a rival lover.[39] When, angered at what he takes to be Elaria's "close intrigues," Cinthio asks Scaramouch to reveal to him the identity of his rival, Scaramouch responds: "*Deceptio visus*, sir: the error of the eyes." With Cinthio's visual mistake echoing the more persistent and prominent visual mistakes that Baliardo makes throughout the play, Scaramouch's Latin commentary operates as an interpretation not only of the immediate scene but of the play as a whole. Consequently, the "groping" in the dark closet that Cinthio performs as he attempts to discern the identity of his companion in the closet functions as an allegory for the groping for information that Doctor Baliardo performs.[40] Furthermore, as an allegory for Baliardo's philosophical investigations, Cinthio's groping in the dark closet becomes an even broader allegory that extends beyond the play to signify contemporary philosophical investigation generally, including that pursued by the likes of Nehemiah Grew and his fellow Royal Society members. I would suggest that the closet parallels the telescope, as Behn portrays both not only to house but also to fool perception in *Emperor.* They are both perceived to disclose hidden information—Baliardo believes the telescope affords him secrets about the world inside the moon; Cinthio believes the closet affords him secrets about the infidelity of his lover Elaria. However, the information that both reveal is false. They accentuate rather than remediate the "error of the eyes."

Thus, like Cavendish, Behn registers serious skepticism about the capacity of empirical prostheses—whether microscopes or telescopes—to enhance philosophers' observational powers; however, for Behn, it is not only perception enhanced by telescopes or dark closets that is prone to mistake. Behn's *Emperor* shows that even observations made in the open air without technological mediation are disposed to inaccuracy. For instance, as Bellemante turns her attention to composing love poetry, she is taken aback when she comes to believe that the poetry has begun to write itself:

> BELLEMANTE: "Tell me, said he, can you resign?"
> —Resign, aye, what shall rhyme to resign? "Tell me, said he"—
> *She lays down the tablets, and walks about. Harlequin peeps from under the table, takes the book, writes in it, and lays it up before she can turn. [She] reads*
> Aye, aye, so it shall be.—
> "Tell me, said he, my Bellemante,
> Will you be kind to your Charmante?"
> *[She] reads those two lines, and is amazed*
> —Ha! Heavens! what's this? I am amazed! And yet I'll venture once more.
> *[She] writes and studies, [and] writes [again]*
> "I blushed, and veiled my wishing eyes."
> *[She] lays down the book, and walks as before*

"Wishing eyes"—
Harlequin writes as before. She turns and takes the tablet
"And answered only with my sighs."—Ha, what is this?
Witchcraft, or some divinity of love? Some cupid sure invisible.
Once more I'll try the charm.[41]

Hidden beneath Bellemante's writing table, Harlequin appears to dupe her into believing that "[w]itchcraft, or some divinity of love" has penned verses to accompany those she has written.[42] True, subsequent lines suggest that she suspects human rather than divine intervention:

—'Tis here again, but where's the hand that writ it? . . .
The little deity that will be seen
But only in his miracles. It cannot be a devil,
For here's no sin nor mischief in all this.[43]

The flirtatious tone of this passage suggests that she presumes that a lover, perhaps Charmante, may be responsible for the "miracles." However, nothing indicates that she suspects Harlequin of the prank. The carpet that adorns the table keeps him from view, shielding his identity from her. Here, then, is another instance of the inaccessibility of empirical facts. Having mistaken Cinthio while hidden in a closet for a secret lover of Elaria, Charmante exclaims: "Was it you, Cinthio? Pox on't, what fools are we, we could not know one another by instinct?"[44] As Behn inserts similar empirical mistakes throughout the play, she suggests that we are all in fact fools or, at least, prone to such foolishness. Consequently, Scaramouch's strategy of hiding from Harlequin by standing right in front of him and pretending to be "a church spout" at the beginning of act 1, scene 3 almost seems legitimate.[45] Behn's *Emperor* suggests that, while we may, like Doctor Baliardo as well as Nehemiah Grew, have a sense that we "take survey of all," our sense perception is, in actuality, woefully limited because it is easily misled.

According to Behn's farce, that stimuli so readily arouse strong responses from observers is largely responsible for the fallibility and, thus, unreliability of our perceptions. It is not only Baliardo who is "ravished" by what he sees. Both Cinthio and Bellemante are, in the examples above, similarly "carried away" and "transported" against their will by sensory stimuli.[46] Indeed, Behn's *Emperor* invites its audience to perceive stimuli to *tickle* observers. In act 1, scene 2 Harlequin comments, "I have somewhere read in an author, either ancient or modern, of a man that laughed to death. I am very ticklish, and am resolved to die that death."[47] He then proceeds to tickle himself: "his head, his ears, his arm-pits, hands, sides, and soles of his feet; making ridiculous cries and noises of laughing several ways, with antic leaps and skips."[48] In featuring this memorable scene early in the play, Behn presents tickling as a metonym for sensation in general. To be tickled is "[t]o touch, or poke (a person) lightly in a sensitive part."[49] Harlequin here serves as a parodic representative of the various other characters in the play, such as Baliardo and Bellemante, who allow and even encourage their sensations to have their way with them, who allow themselves to be "touched . . . in a sensitive part" by what they observe. Baliardo tickles himself as he engages in telescopic observations. Bellemante tickles herself as she allows herself to believe in her automatic writing tablet. Scaramouch emphasizes this theme of tickling when in act 2, scene 3 he observes, in reference to Doctor Baliardo's susceptibility to the story of the Emperor of the Moon, "This madness is a pretty sort of a pleasant disease, when it tickles but in one vein. Why, here's my master now: as great a scholar, as grave and wise a man, in all argument and discourse, as can be met with; yet name but the moon, and he runs into ridicule, and grows as mad as the wind."[50] Behn suggests that to sense is to be tickled. The tickle of sensation is pleasurable when it remains moderate but becomes overwhelming, and may even induce madness, when it is allowed to run amok.

I would contend that Behn's portrayal of the tickle of sensation is informed by Creech's translation of book 4 of *De Rerum Natura*. There, Lucretius as translated by Creech reflects on the "tickling Art" of images that provoke love:

> Love rises then, when from a *beauteous* face
> Some *pleasing* forms provoke us to embrace
> Those *Bawds to Lust*, when with a *tickling* Art,
> They gather turgent *seed* from every part;
> And then provoke it, then rise *fierce* desires;
> The *Lover* burns with strong, but pleasing fires
> Those often are pursu'd by *following* Care,
> Distracting thoughts, and often deep despair.[51]

Creech's translation here suggests that we cannot help loving and/or desiring when "pleasing forms" or images "tickle" us in this way. According to Trull, Lucretius via Creech indicates in these lines that "[t]he soul has no means of resistance against the incursion of images; they move the soul by direct contact, the '*tickling* Art' that irresistibly gathers 'seed' from the body into the genitals."[52] Incorporating the notion of tickling sensations into *The Emperor of the Moon*, Behn emphasizes the greatly stimulating and even irresistible appeal of what we perceive. We cannot help but be seduced by sensory stimuli.

By simultaneously insisting that humans are fundamentally prone to being stimulated (tickled or charmed) by what they sense and that Baconian modesty fails to enable the philosopher to close himself off to the stimulation of sensation, Behn's *Emperor* offers a dramatic depiction of how challenging it is to be a successful new philosopher. According to Behn's satire, the cards are significantly stacked against Baconian empiricists. However, she nevertheless offers a solution to the philosophical challenges she identifies. Bacon presented modesty as a key method by which to overcome or at least resist the difficulties he acknowledged; the practical solution Behn suggests could hardly be more different than that proposed by Bacon. Whereas Bacon advises philosophers to purify themselves of matter and, thus, to establish distance between themselves and the objects of their investigation, Behn insists that philosophers acknowledge rather than deny their material connection or entanglement with the objects of their investigation. According to Behn, in doing so, they become more capable of discernment and, thus, more successful in their empiricism. For Bacon, it is in denying their material entanglement that philosophers ensures the legitimacy of their practice; for Behn, it is in recognizing their material entanglement that they do so. The character in *The Emperor of the Moon* who offers a positive example of this practice is one whom scholars tend to overlook: Mopsophil. Though Mopsophil is a secondary character and, therefore, primarily active in the subplot rather than the main plot of the play, I would propose she nonetheless presents an important alternative to the other varieties of observer that Behn presents. Mopsophil shows herself to be a sensitive witness, and it is as a result of her openness to feeling, rather than in spite of it, that she is capable of perceiving hidden details where others tend to remain oblivious to them.

Her effectiveness as an observer becomes most apparent in the final scene of the play. In this scene, Behn juxtaposes the main plot with the subplot and emphasizes their contrast. In the main plot, modest Doctor Baliardo is duped into approving of his daughter's and niece's marriage to their lovers, having been ludicrously convinced that they are in fact marrying the Emperor of the Moon and the Prince of Thunderland, both of whom have supposedly just descended to earth in a chariot from the moon. In the subplot, however, the parallel ruse concocted by Scaramouch and Harlequin only partially succeeds:

KEPLER: What are you?
SCARAMOUCH: Two neighbouring princes to your vast dominion.
HARLEQUIN: Knights of the sun, our honourable titles,
And fight for that fair mortal, Mopsophil.
MOPSOPHIL: [*aside*] Bless us! My two precious lovers, I'll warrant; well, I had better take up with one of them, than lie alone tonight.[53]

Disguised as "Knights of the sun," Scaramouch and Harlequin proceed to "try [their] titles by the dint of lance" with the emperor as "arbitrator."[54] Scaramouch is pronounced "victor," and the Doctor, having fallen for yet another ruse, subsequently offers Mopsophil to the prince (aka Scaramouch) "as the reward of [his] undoubted valour."[55] Mopsophil, however, sees through the joke rather than being deluded by it. Indeed, she is even capable of accurately identifying Scaramouch and Harlequin, in spite of their wearing "helmets" that hide their faces.

This contrast between Mopsophil and the Doctor is striking. The Doctor's perception of his own "refinement" has helped make him into a fool by causing him to fall prey to deception. Mopsophil, on the other hand, appears in no way refined, nor does she perceive herself as such. In fact, here as elsewhere in the play, she appears as something of a libertine.[56] Motivated by a desire to have company in bed and, thus, presumably by a desire for sexual pleasure, she seems perfectly happy to "take up with" either of her "precious lovers." As we saw above, the Doctor believes that his "strength of virtue" and "purity" enable him to see what others cannot and ultimately, thus, to become privy to the goings on in the moon. Mopsophil has no such faith. Indeed, I would suggest a correlation between her relaxed acceptance of her place in the material world and her uniquely robust capacity for discernment. According to Behn's portrayal, Mopsophil's acknowledgment of the fact of her own sexual and, thus, material being primes her to be a sensitive witness—an observer attuned, and palpably so, to empirical facts even when they lie beyond the surface.

"To the Unknown Daphnis on his Excellent Translation of Lucretius"

Behn also features the practice of sensitive witnessing in the poem she writes to Creech in celebration of his translation of *De Rerum Natura*. This portrayal helps illustrate that, as is the case with Hutchinson and Cavendish, Lucretian Epicureanism significantly informs her impulse, throughout various of her works, to promote sensitive witnessing as an alternative to Baconian modest witnessing. Behn begins "To the Unknown Daphnis on his Excellent Translation of Lucretius" with a gendered contrast between "those that sing thy mighty Praises Loud" and herself who possesses a "humbler Muse."[57] The lines that follow elaborate this contrast:

> Inspir'd by Thy vast Flights of Verse
> Methinks I should some wonderous thing Reherse
> Worthy Divine Lucretius, and Diviner You!
> But I of feebler Seeds design'd,
> While the slow moveing Atoms strove
> With Careless Heed to Form my Mind,
> Compos'd it all of softer Love:
> In Gentle Numbers all my Songs are drest:
> And when I would Thy Glories sing,
> What in Strong Manly Verse should be exprest
> Turns all to Womanish Tenderness within;
> Whilst that which Admiration does Inspire,
> In other Souls, kindles in Mine a Fire.[58]

With this opening, Behn assumes a humble stance as she situates herself in opposition to poets, such as Creech, who pursue "vast Flights" and, thus, pen "Strong Manly Verse." Inspired by "Admiration," such masculine poets depict "wonderous things"; in contrast, her soul "kindles" with "Fire." She is inspired by "Love" and desire. Behn indicates that, whereas masculine poets are powerful and energetic, her female sex makes her soft, tender, and delicate: she has been "design'd" with "feebler Seeds" and "slow[er] moveing atoms" than those used to create her masculine counterparts.

Though Behn introduces this poem with an expression of feminine humility and submission, she complicates such humility in the turn the poem makes at line seventeen. In this passage she indicates that, in spite of masculine poets' advantages, what she gives Creech in this verse in fact surpasses what they give:

> Let them Admire thee on—whilst I this newer way
> Pay thee yet more than They,
> For more I ow, since thou hast taught Me more,
> Than all the Mighty *Bards* that went before.[59]

Behn here claims to pursue a "newer way" of poetry whose yield is, at least for Creech, more substantial than that afforded by the path that her masculine counterparts pursue. The implication is that the rest of the poem embodies the new way that these lines establish and thus provides an example of an ode in praise of a fellow poet that is successful in spite of (really, by virtue of) its femininity. Given the portrayal of femininity that Behn provides in the introduction of the poem, as quoted above, readers are invited to expect that the rest of the poem will be "soft" in volume and in texture, "gentle" and "tender." We are also prepared to observe Behn's love and desire ("fire") for Creech. Recognizing her femininity, Behn presumes a palpable physical and emotional intimacy with Creech and his translation. Whereas her masculine counterparts "sing [his] mighty Praises Loud," and thus boldly and publicly assert themselves, she intends a quieter performance in which she will connect with, and feel for, Creech affectively and in a more personal and embodied fashion. In other words, Behn casts herself as a sensitive witness in "To the Unknown Daphnis" and invites her readers to appreciate the distinctive poetic style that her practice of sensitive witnessing produces.

Typically distinguished by abstraction, formality, ceremoniousness, and an elevated style, the ode is a surprising choice of genre with which to convey such a personal and emotional practice. In fact, I would suggest that the poem operates as a parody of the traditional ode. As Alvin Snider observes, "[i]n Behn's opening salvo, we find the speaker physicalizing lofty sentiment, conjoining the libertine and the libidinal, and dramatizing how the irresistible pull of pleasure produces a kind of automatism in the nominally free subject. The poem provides a site for the representation of the body as vulnerable to inspiration and passion, as subject to internal and external forces that reduce the autonomy of individual agents."[60] In "To the Unknown Daphnis," like in *The Emperor of the Moon*, Behn portrays the sensitive witness as a libertine, and as she does so her ode operates as a mock ode with the genre's conventional lofty celebration becoming transformed into a distinctively earthly and material expression of appreciation. Behn writes of "souls," but she insists upon their association, and even continuity, with the erotic body. Thus denying the fundamental opposition between spirit and matter, she continues to undermine the binaristic structure she uses to introduce and contextualize the poem. The soul reveals itself as material; her feminine verse reveals itself to be worth more than the verse of her masculine counterparts; the lofty and public ode reveals itself to be compatible with personal emotion and even erotic desire.

Behn's "Muse" might be "humbler," but the "Fire" that invigorates her poetry is nothing to be sniffed at. Expressed by Behn, "Womanish Tenderness" comes to warrant the reader's respect. Admiring Creech, she asserts that she too is worthy of admiration. The "tender" poetry of the sensitive witness is at least as, if not more, valuable than the "Strong Manly Verse" composed by "the Croud of those that [would] sing thy mighty Praises Loud." Indeed, it expresses androgyny as it embodies both femininity (in its soft humility) and masculinity (in its power).

Behn credits Creech with helping her overcome her presumed feminine inferiority and thereby surpass the binary and hierarchy of gender. In fact, she indicates that the main reason she is so enamored with his translation of Lucretius is that it has allowed her to gain access to knowledge and, thus, power that are otherwise, because of her sex, beyond her reach: "Thou by this *Translation* dost advance / Our Knowledge from the State of Ignorance; / And Equallst Us to Man! Oh how shall We / Enough Adore, or Sacrifice enough to Thee!"[61] She explains, "Till now I curst my *Sex* and *Education*, / And more the scanted Customs of the Nation, / Permitting not the Female Sex to tread / The Mighty Paths of Learned *Heroes* Dead."[62] According to these lines, with Creech's translation of Lucretius, women have become capable of emulating legendary "learned heroes" and thus defying their "education" and the national "customs" with which their education coincides. She goes on to observe:

> The Godlike *Virgil* and Great *Homers* Muse
> Like Divine Mysteries are conceal'd from us,
> We are forbid all grateful Theams,
> No ravishing Thoughts approach our Ear;
> The Fulsom Gingle of the Times
> Is all we are allow'd to Understand, or Hear.[63]

Behn here bemoans women's lack of access to poetic sublimity ("ravishing Thoughts"). With his translation of Lucretius, though, Creech has remediated this lack and has done so in a way that translates the transporting nature of Lucretius's poetry into a style that is distinctively appealing to the feminine sensitive witness but that nevertheless preserves its masculinity and sublimity. As the passage quoted above in which Behn analyzes "[t]he Mystick Terms of [Creech's] Rough *Philosophy*" illustrates, she not only claims androgyny for women but also attributes it to Creech. She indicates that, in his version of Lucretius's *De Rerum Natura*, Creech makes masculinity compatible with femininity: "Rough" philosophy manifests in feminine form ("soft and Gay a Dress"), its "Terms" are both instructive to the mind and charming to the "Sense," it joins "Eloquence" and "Reason" with "Fancy" and playfulness. His verse is both "soft" and "strong," "wanton" and yet resolute ("no hard notion meets, or stops its way").[64] According to Behn's portrayal, as Creech's translation embodies such gendered contradictions, it surpasses gender binaries much as her ode does. Behn therefore suggests that his Lucretius sets an important precedent for her and provides a valuable example.

A Discovery of New Worlds

Behn may also be seen to celebrate philosophy that surpasses gender binaries by incorporating softness and sensitivity in her translation of Fontenelle's 1686 *Entretiens sur la pluralité des mondes*. According to Behn, Fontenelle's main goal with this work was "to treat . . . Natural Philosophy in a more familiar Way than any other hath done, and to make every body understand him."[65] To that end, he composes the book as a series of conversations between a marquiese and a philosopher who, inspired by Copernicus, explore the possibilities of extraterrestrial life as they stroll through the Marquiese's garden at night. A desire to entertain a wide audience is ostensibly a prime motivator for Behn in her composition of both *The Emperor of the Moon* and *A Discovery of New Worlds*. As she acknowledges in the latter's preface, "The General Applause this little Book of the Plurality of Worlds has met with, both in France and England in the Original, made me attempt to translate it into English."[66] She tells William, Earl of Drumlangrig in her dedicatory epistle that she believes it "has something in it out of the way of ordinary Wit," in spite of the fact that she views it as "but a trifle."[67]

However, neither her focus on her audience's pleasure nor her conventional performance of feminine deference should be allowed to detract from the translation's philosophical significance. Claiming authority for the work, Behn also claims authority for the variety of sensitive witnessing the work promotes. Behn may ask the Earl to "pardon" her translation "[i]f it is not done with that exactness it merits" for she is a "Woman, who is not supposed to be well versed in the Terms of Philosophy, being but a new beginner in that Science."[68] Nevertheless, in publishing the work, Behn offers a strong assertion of her own philosophical expertise.[69] The confidence and assertiveness with which she engages both astronomical and biblical controversies in her preface offset her performance of deference. In fact, she directly challenges Fontenelle himself:

> How well [Fontenelle] hath performed his Undertaking you will best judge when you have perused the Book: But if you would know beforehand my Thoughts, I must tell you freely, he hath failed in his Design; for endeavouring to render this part of Natural Philosophy familiar, he hath turned it into Ridicule; he hath pushed his wild Notion of the Plurality of Worlds to that heighth of Extravagancy, that he most certainly will confound those Readers, who have not Judgment and Wit to distinguish between what is truly solid (or, at least, probable) and what is trifling and airy: and there is no less Skill and Understanding required in this, than in comprehending the whole Subject he treats of.[70]

The assuredness with which she criticizes Fontenelle's work in this passage is striking, and as she proceeds to present translations of Fontenelle's author's preface and dedicatory epistle, she only further affirms her authority. Indeed, Fontenelle himself attests to the philosophical capacity of women: "In this Discourse I have introduced a fair Lady to be instructed in Philosophy . . . imagining, by this Fiction, I shall render my Work more agreeable, and to encourage the fair Sex . . . by the Example of a Lady who had no supernatural Character, and who never goes beyond the Bounds of a Person who has no Tincture of Learning, and yet Understands all that is told her."[71]

In a vein similar to *The Emperor of the Moon*, *A Discovery of New Worlds* satirizes philosophical modesty while promoting sensitive witnessing as a far preferable alternative. In a critical portrayal of the modest witness, Fontenelle's philosopher-protagonist asserts, "the same inclination that makes us endeavour to have the first place in a Ceremony, prevails with a Philosopher in composing his System, to place himself in the Centre of the World, if he can: He is proud to fansie all things made for himself; and without reflexion, flatters his Senses with this Opinion, which consists purely in Speculation." [72] Here, the modest philosopher's desire to set himself apart is revealed as solipsism, "pride," and self-"flattery." Fontenelle's philosopher-protagonist and "fair Lady" model a different, sensitive rather than aloof, approach to knowledge-making. It is primarily the Marquiese who sets this style and tone for their practice. From the beginning of the work, she is portrayed as a simultaneously discerning and sensuously receptive character. *Discovery* opens with a discussion about the relative beauties of the night. Insisting on the superiority of night's appeal, the Marquiese asserts, "you must acknowledge that the brightest day that ever you saw could never have engaged you in so agreeable an Ecstasie, as you were just now like to have faln into by the powerful attractions of this Night."[73] She goes on to observe, "those fair Beauties that so resemble the day, produce not those soft Effects of the other. How comes it, that Loves who are the best Judges of what is pleasing and touching do always address themselves to the Night, in all their Songs and Elegies?"[74] As these lines show, from the beginning of *A Discovery of New Worlds*, the Marquiese is depicted as an admirer of "what is pleasing and touching" and associated with "soft Effects." Such sensuousness and softness set the tone for the rest of the work.

Fontenelle directly answers detractors who say that the female "*Sex must needs be wanting, in those Perfections which adorn ours, because they do not read so much.*"[75] He offers his Madam la Marquiese as an "Example of a Lady who . . . retains all the Notions of *Tourbillions* and Worlds, without Confusion."[76] It may sound like self-deprecation when the Marquiese depicts herself as equipped with "Ignorance" while her philosopher-guide is equipped with "Systems."[77] However, readers are invited to view her "ignorance" as an asset rather than a hindrance, for it is precisely what allows her to be a good student of philosophy. Her lack of prior knowledge and her humility make her receptive and, therefore, responsive to what the philosopher presents. She is sensitive to the impressions of the cosmos with which the philosopher presents her. His presentations affect her and, thus, shift her perspective. At the same time, she is never fully swept away because, much like Mopsophil in *The Emperor of the Moon*, she is fully aware of how easy it is to be swayed. She maintains an openness to the philosopher's presentations but is simultaneously persistent with her evaluation of their plausibility and persuasiveness. The Marquiese may be associated with softness and sensitivity, but in the words of the philosopher, her "Judgment and Understanding is solid and penetrating."[78]

Indeed the philosopher and the Marquiese regularly determine philosophical truths collaboratively. She points out the inconsistencies in the theories he presents and, together, they determine what the most plausible theory is through a dialogic analysis of what informs the theories. For instance, at the conclusion of their evaluation of Copernicus's theory of heliocentrism, the philosopher explains, "At last, we resolv'd to hold our selves to the Opinion of *Copernicus*, which is more uniform, and more agreeable, without the least mixture of Prejudice; and, indeed, its Simplicity perswades as much as its Boldness pleases."[79] This and other similar passages suggest that it is not only the Marquiese who practices sensitive witnessing in this work. The philosopher himself appears to do so as he matches the Marquiese's responsiveness and openness with his own. Here as in *The Emperor of the Moon*, material sensitivity is presented as not only compatible with, but in fact supportive of, philosophical discernment. In the few places where the Marquiese's practice of sensitive witnessing is portrayed to have come up short, it is when she has gotten ahead of herself and come to a conclusion too soon. In other words, it is when her sensitivity has lapsed for the sake of a hasty resolution. This happens when the philosopher presents her with the idea that the moon is inhabited. At the beginning of the Third Night, she explains "You made me believe there were Inhabitants in the Moon; I have overcome all the Difficulties my Reason suggested to me against that Opinion, and now I am resolv'd I will believe it." The philosopher responds thus: "You go too fast, . . . Madam; one ought to give but one half of one's Thoughts and Belief to Opinions of this nature, reserving the other half free for receiving the contrary Opinion, if there be occasion." And then, the Marquiese insists, "I am not to be deluded . . . with fair Words; let us come to the subject matter in debate."[80] The give-and-take of the calm and respectful form of "debate" that the philosopher and the Marquiese pursue throughout the entire work is an ideal form of communication for the sensitive witness. Both episodic and recursive, extended debate allows the sensitive witness to adjust their point of view to accommodate new impressions while maintaining a regular commitment to reasonable evaluation that refuses to be too easily seduced by "fair words."

Oroonoko

Though *The Emperor of the Moon* and *A Discovery of New Worlds* emphasize, albeit in different ways, the philosophical value Behn attributed to sensitive witnessing, no study of witnessing in the writings of Aphra Behn would be complete without an analysis of *Oroonoko*. It is in this work that the practice achieves its fullest expression. *Oroonoko* was written within a year of the other works and was published in the same year as her translation of Fontenelle. All three works register the influence of Epicurean materialism and, I argue, draw inspiration from it as they promote methods of sensitive witnessing while challenging the legitimacy and value of the Baconian modest witness.[81] From the beginning, witnessing is a central theme in *Oroonoko*. Related terms recur throughout the work, starting as early as the second paragraph, which begins thus: "I was my self an Eye-Witness to a great part, of what you will find here set down; and what I cou'd not be Witness of, I receiv'd from the Mouth of the chief Actor in this History, the *Hero* himself, who gave us the whole Transactions of his Youth."[82] Behn's narrator establishes the veracity and, thus, authority of her story by asserting her effectiveness as a "witness." According to this passage, her practice of witnessing takes two main forms: (1) direct visual perception ("eye-witnessing"), and (2) perception mediated through Oroonoko's narration of his history. Behn's narrator claims to have attained a comprehensive impression of the hero's history by virtue of her practice of witnessing: "[T]hough I shall omit, for Brevity's sake, a thousand little Accidents of his Life, which, however pleasant to us, where History was scarce, and Adventures very rare, yet might prove tedious and heavy to my Reader, in a World where he finds Diversions for every Minute, new and strange: But we who were perfectly charm'd with the Character of this great Man, were curious to gather every Circumstance of his Life."[83] Though she here acknowledges that, so as not to be "tedious and heavy," she has been selective about what she has shared with her reader, she insists that the selection she provides is informed by a complete view. She has been "curious to gather every Circumstance of his Life" and claims to have been successful in having done so.

That Behn portrays her narrative persona as having been "charm'd" by Oroonoko in the above quote attests to the influence of Lucretian Epicureanism on the distinctive variety of witnessing that she adopts and portrays in the work. As we observed above, across various of Behn's writings, the term *charm* signals "the involuntary or mechanical power of attraction" poeticized by Lucretius.[84] In using the term to signify her narrative persona's relationship to Oroonoko, she emphasizes their intimacy. Moreover, to recognize the Epicurean valence of the term is to appreciate the profundity of such intimacy. It is not merely rhetorical, instrumental, intellectual, or even emotional. Using Lucretian Epicureanism as an interpretive guide, which the term *charm* invites one to do, Behn's reader discovers the materiality of their intimacy. It goes as deep as their bodies' atoms. It as if the narrator is as enthralled with Oroonoko as a lover would be. The term *charm* recurs throughout *Oroonoko* much as it does, as Ballaster has shown, throughout Behn's amatory poetry. When other characters are charmed, it frequently connotes erotic desire. For instance, when Oroonoko first encounters Imoinda, he is "charmed" by "the Beauty of this fair Queen of Night."[85] Similarly, when the old King meets Imoinda, he perceives her "as the most charming he had ever possess'd in all the long Race of his numerous Years."[86] Behn's narrator is regularly effusive in her expression of admiration for Oroonoko. Characterizing him as a "[w]onder of all that World," she declares that he "was adorn'd with a native Beauty so transcending all those of his gloomy Race, that he strook an Awe and Reverence even in those that knew not his Quality; as he did in me, who beheld him with Surprize and Wonder."[87] The narrator is enraptured with Oroonoko—not only with his "Beauty" but also his "Valour and Magnanimity."[88] Overcome by both his physical splendor and his "Character," Behn's narrator finds Oroonoko utterly irresistible.[89]

Numerous scholars have offered extensive treatments of Behn's hyperbolic celebration of Oroonoko. Claiming that Behn composes Oroonoko as a parallel for herself, Catherine Gallagher asserts that the novella "associates the blackness of the character . . . with authorial transcendence."[90] Thus, in Gallagher's influential analysis, "Oroonoko resembles the mystical body of the text; his blackness is a luminous emanation of authorship, which gleams forth from multiple inscriptions."[91] Whereas Gallagher shows how Behn, in a sense, domesticates Oroonoko, employing him as a figure for herself, Ramesh Mallipeddi instead emphasizes the character's exoticism.[92] Analyzing Oroonoko through Hayden White's theory of the "Noble Savage," Mallipeddi argues that his "body is fetishized as an exotic commodity" within the novella.[93] As Laura Brown asserts, crucial to the admiration Behn expresses for the "royal slave," and the "sentimental identification" that such admiration enables, is her assimilation of him to European ideals.[94] Behn's depiction of him is an "act of absolute identity."[95] As Brown proceeds to explain, Oroonoko is "not only a natural European and aristocrat, but a natural neoclassicist and Royalist as well, an absurdity generated by the desire for an intimate identification with the 'royal slave.' "[96]

To recognize the novella's affiliation with Lucretian Epicureanism, though, is to reveal another important framework for analyzing the admiration of, and sentimental identification with, Oroonoko that Behn's narrator expresses. Behn's Epicureanism helps shape, and perhaps even inspire, the enthusiasm and intimacy with which she conveys her novella's hero. Epicurean materialism precludes essential differences—or even material distance—between humans. If bodies are porous and atoms pass between them, then where does Oroonoko's body end and Behn's narrator's begin? Transcorporeal bodies invalidate essentialist racial, ethnic, or national distinctions, revealing them to be *un*natural and imaginary. Interpreted through the framework of Epicureanism, then, the identification of the narrator with the hero appears founded on a perception of their fundamental material entanglement. Their unity, not their difference, is primary. Distinguished by chaos—what Lucretius terms "the clinamen" or "swerve"—and beyond perception, such material unity verges on ineffability. The following passage by Lucretius on the clinamen attests to the difficulty of pinning down atomic chaos:

> While atoms move by their own weight straight down
> Through the empty void, at quite uncertain times
> And uncertain places they swerve slightly from their course.
> You might call it no more than a mere change of motion.
> If this did not occur, then all of them
> Would fall like drops of rain down through the void.
> There would be no collisions, no impacts
> Of atoms upon atom, so that nature
> Would never have created anything.[97]

Lucretius feels compelled to resort to describing what atoms are *not* in order to point to what atoms are and how they move. Unpredictable ("uncertain") and imperceptible, atoms evade one's desire to describe them. We might therefore further expand our understanding of the awkwardness and "absurdity" of Behn's Europeanization of Oroonoko by recognizing it as a consequence of attempting to depict discursively ineffable Epicurean materiality. Such depiction is prone to be a little clumsy or at least approximate.

The distinctive practice of witnessing that Behn portrays in *Oroonoko* is also reflective of Epicurean principles. As Rebekah Mitsein has shown, *Oroonoko* "mimics" contemporary travel reports and "scrutinizes the ways in which geographical representations became palimpsests of eyewitness details, oral accounts, and narrative evidence in the early Enlightenment."[98] Significantly influenced by empirical practices endorsed and employed by the Royal Society, such reports performed, and gave the impression of having been informed by, modest witnessing. In fact, Robert Boyle published a guide in 1692 to help travelers employ philosophical modesty in their voyages and the accounts they composed to document them. Boyle explains the motivation behind, and goal of, his guide thus:

> Considering the great Improvements, that have of late been made of Natural History (the only sure Foundation of Natural Philosophy,) by the Travels of Gentlemen, Seamen, and others; And the gread [*sic*] Disadvantage many Ingenious Men are at in their Travels, by reason they know not before-hand, what things they are to inform themselves of in every

> Country they come to, or by what Method they may make Enquiries about things to be known there, I thought it would not be unacceptable to such, to have Directions in General, relating to all, and also in Particular, relating to Particular Countries, in as little Bounds as possible, presented to their View.[99]

In this introduction, Boyle acknowledges how much travel reports have to contribute to the pursuit of "Natural History (the only sure Foundation of Natural Philosophy)." He seeks to bring travel pursuits and documentation into line with accepted empirical practices in order to help maximize such contributions.

Most relevant to our study of *Oroonoko* are Boyle's recommendations for observing and depicting "inhabitants": "The Inhabitants themselves are to be consider'd, both Natives and Strangers, that have been long settled there; particularly their Stature, Shape, Features, Strength, Ingenuity, Dyet, Inclination, that seem not due to Education. As to their Women, their Fruitfulness or Barrenness, their easie or hard Labour, with their Exercises and Dyet; the Diseases both Men and Women are subject to, peculiar to themselves, compared with their Dyet, Air, *&c.* that do influence them."[100] Here, Boyle may be seen to promote, and implicitly emphasize the importance of, the detachment of the observer from the observed. Though interpersonal exchanges would surely be necessary to determine some of these details, Boyle ignores such exchanges and gives the impression that the details are being witnessed from afar. "The Inhabitants" are to be "consider'd"—not encountered. In fact, it is as if he advises the traveler to observe people in precisely the same way as he advises them to observe the landscape:

> What are its Dimensions, Situation, East, West, South or North, its Figure, its Plains, Hills or Valleys, their Extent, the highth of the Hills, either in respect of the neighbouring Valleys, or the Level of the Sea; as also whether the Mountains lye scatter'd or in Ridges, and whether those run North or South, East or West, *&c.* What Promontories, Fiery or Smoaking Hills, *&c.* the Country has or hath not; whether subject to Earthquakes or not. Whether the Country is coherent, or much broken into Islands.[101]

In accordance with Francis Bacon's method of "purification" through masculine modesty, Boyle prompts the "Ingenious" traveler to maintain a consistent aloofness throughout his voyages and, thus, to subject his fellow humans to the same detached empirical evaluation as he employs in his observations of nonsentient features of foreign lands. As David Livingstone explains, explorers needed to be "properly trained eyewitnesses" for their reports to be useful to natural history and, thus, natural philosophy.[102] This training involved learning how to maintain a physical and psychological separation of oneself from the landscapes, objects, and people one encountered in one's travels. Such separation was instrumental in appearing to ensure the neutrality and transparency that were defining features of Baconian modesty.

Boyle's guide merely helped promote what had, by the time of its publication, become a common characteristic of travel narratives. Baconian modesty runs throughout seventeenth-century explorers' reports on their experiences in foreign lands. See for instance Jobson's depiction of Ethiopians:

> The people, who are Lords, and Commaunders of this country, and professe themselues the naturall Inhabitants, are perfectly blacke, both men and women. The men for their parts, do liue a most idle kinde of life, imploying themselues (I meane the greater part) to no kinde of trade nor exercise, except it be onely some two moneths of the yeare, which is in tilling, and bringing home their countrey corne, and graine, wherein the preseruation of their liues consists, and in that time their labour is sore, as when I come to shew the manner, you may easily conceiue; All other times of the yeare, they liue wandring vp and downe, from one to an other, hauing little vnderstanding, either to hunt in the woods, or fish in the waters; notwithstanding, both the one and the other, in their kindes, are infinitely replenished, that to their very doores wilde beasts doe resort, and about their houses in euery corner, abundance of *Ginny* hennes, and excellent partridges.[103]

The anonymously published *The Golden Coast* (1665) employs a similarly modest method of witnessing. See for instance: "The Husbands and Wives neither eat nor lye together but on their Tuesdaies, when they spread a Mat upon the ground, and lay stools under their heads with a fire at the end of the bed to warme the soles of their feet."[104] The privacy and intimacy of this detail is offset by the observer's empirical modesty. His aloofness and the neutrality that it achieves counter the sexual voyeurism that the information implies. Himself purified of matter, the explorer is there but not there.[105]

Though the seventeenth-century travel narrative genre provides an important context for and influence on Behn's *Oroonoko*, her portrayal of witnessing significantly deviates from and challenges key conventions of the genre. Behn regularly portrays "modesty" in a favorable light in *Oroonoko*, but the modesty she admires is not the same as that promoted by Bacon and his followers. As Tita Chico explains, the Baconian "modest witness is a masculine figure of authority, gentility, and privilege, admired for his morality and knowledgeability and, just as notable, distinguished from women and laboring men. He embodies, paradoxically, a privileged absence, a model of spectatorship that conveys authority through its claim to be free from the limitations of self-interest."[106] In the words of Donna Haraway, he is perceived to be "on the side of the objects" and, thus, "transparent."[107] Whereas Bacon and his followers invested in modesty that signified masculine transparency, Behn promotes modesty aligned instead with feminine *softness* and *delicacy*. The character to whom Behn most often attaches the term *modest* is Imoinda. For instance, Behn depicts Oroonoko's first meeting with Imoinda thus: "he was infinitely surpriz'd at the Beauty of this fair Queen of Night, whose Face and Person was so exceeding all he had ever beheld, that lovely *Modesty* with which she receiv'd him, that Softness in her Look, and Sighs, upon the melancholy Occasion."[108] Trefry's depiction of her as "Clemene," "the "fair Slave," closely corresponds to Oroonoko's:

> She is adorn'd with the most Graceful Modesty that ever beautifyed Youth; the softest Sigher—that if she were capable of Love, one would swear she languish'd for some absent happy Man; and so retir'd, as if she fear'd a Rape even from the God of Day; or that the Breezes would steal Kisses from her delicate Mouth. . . . But oh! she disarms me, with that Modesty and Weeping so tender and so moving.[109]

To be "modest" is, according to Behn's portrayal, to be soft, "tender," vulnerable, susceptible to impressions both physical and emotional. In other words, for Behn, to be modest is to be *sensitive*.

Considering her passivity, submissiveness, and limited dialogue, it is tempting to think that Imoinda's role and importance within *Oroonoko* is highly circumscribed. Indeed, we might even feel inclined to criticize Behn for the limited role Imoinda plays within the novella, particularly considering the strength and strong voice with which the author imbues European female characters in other of her works—*The Rover* being only one of many examples. However, I would argue that, as a figure for feminine modesty, Imoinda is far more central to *Oroonoko* than she superficially seems. In fact, I would suggest that Imoinda represents precisely the type of witness that—with her portrayal not only of Imoinda but also of the novella's narrator—Behn seeks to embody, promote, and have her readers emulate. Like Imoinda, Behn's narrative persona is distinguished by feminine modesty. She too is highly sensitive. She demonstrates such sensitivity throughout the novella, but its most overt expression appears at the end, after the surgeon has indicated that Oroonoko's death is imminent. Of Oroonoko's maimed body, Behn's narrator recounts, "the Sight was gashly; his Discourse was sad; and the earthly Smell about him so strong, that I was perswaded to leave the Place for some time (being my self but Sickly, and very apt to fall into Fits of dangerous Illness upon any extraordinary Melancholy)."[110] Here, Behn points to her narrator's sensitivity with the word "sickly" and particularly emphasizes her susceptibility to "Melancholy." So sensitive is the narrator to "the sight" of Oroonoko's wounded body that she must take a break from her eye-witnessing by "leaving the place for some time."

Though this passage on its own may appear to indicate the limit of the narrator's capacity to serve as a witness, considered in the broader context of the novella, it in fact helps to reinforce her special skills as a witness. From the work's beginning, Behn suggests that the only reason we are able to become privy to Oroonoko's story is because of the narrator's feminine gentleness and sensitivity to his trials—traits that mirror Imoinda's. In the second paragraph of the novella, Behn's narrator explains, "I was my self an Eye-Witness to a great part, of what you will find here set down; and what I could not be Witness of, I received from the Mouth of the chief Actor in this History, the *Hero* himself, who gave us the whole Transactions of his Youth."[111] Later she goes on to explain that she has been able to gain special access to Oroonoko's story because she has a uniquely intimate camaraderie with him: "he lik'd the Company of us Women much above the Men . . . we had all the Liberty of Speech with him, especially my self, whom he call'd his *Great Mistress*; and indeed my Word wou'd go a great way with him."[112] Behn's narrator proceeds to describe how she works to "soften" her words when, on one occasion, she has found them to trigger "resent[ment]" in Oroonoko.[113] Softness—susceptibility, sensitivity, responsiveness—distinguishes the practice of witnessing that Behn models in *Oroonoko*, and as these passages suggest, it enables her narrator to observe what would be lost on (or intentionally hidden from) an aloof masculine witness. Her "Word . . . goes a great way with him" because she allows his words to go a great way with her. She is highly affected by Oroonoko's speech and his experiences. This allows her great "Liberty" not only in "Speech" but also in her observations of Oroonoko. Reciprocity thus appears to characterize the relationship between Behn's narrator and Oroonoko. Such reciprocity helps explain why she has intimate access to Oroonoko's experience but not to that of his fellow enslaved. Behn's narrator refuses to stake claim to knowledge that has not been informed by sensitive witnessing and the reciprocal dynamic that such witnessing entails. Her understanding and knowledge are predicated on intimacy, closeness, and relationality. Therefore, what may superficially appear to be a failure of imagination on the part of Behn's narrator in fact reflects her refusal to presume that she knows more than her position allows. Behn implies that her narrator simply cannot effectively describe the plight of other enslaved individuals because she has not yet had an opportunity to observe their experiences with sensitivity and tenderness.

Of course, the reciprocity between Behn's narrator and Oroonoko is belied by the power imbalance between them. Indeed, this scene in which Behn's narrator claims that her "Word wou'd go a great way with him" confronts us with how sensitive witnessing could be put to sinister uses, though as we will see, Behn ultimately criticizes such abuse of the practice.[114] Suvir Kaul has suggested that *Oroonoko* conveys "what Mary Louise Pratt calls an idealized 'drama of reciprocity,'" and this scene offers a particularly significant expression of this drama.[115] As Kaul explains, such "brief enactment[s] of reciprocity" are necessarily supplanted by "a quick and reassuring reassertion of superiority."[116] This is precisely what occurs in this scene. As Laura Doyle indicates, "following the narrator's distancing herself from the society of common men, she uses the motions of a sentimental sensibility to re-configure her alliances so that she moves from an 'us' aligned with Oroonoko to an 'us' aligned against him—joining, that is, the new colonial racial order despite all of her feminine sympathy with Oroonoko's nobility."[117] After having just established her mutual intimacy with Oroonoko—even claiming his "love"—by separating herself and her fellow female English travelers from the male English colonists, Behn's narrator proceeds to shift her alliance so that she becomes identified with the male colonists after all. She explains, "I took it Ill he shou'd Suspect we wou'd break our Words with him, and not permit both him and *Clemene* to return to his own Kingdom, which was not so long a way, but when he was once on his Voyage he wou'd quickly arrive there."[118] Obviously, it would be the male colonists, not the English women, who would have such decision-making power. Therefore, the "we" in this passage must include both English men and women. Furthermore, whereas in the earlier section of the scene when Behn's "we" includes only women, Oroonoko is portrayed as a companion, in this latter stage as the "we" expands to include both men and women, he is portrayed as a slave, subject to the English colonists' orders and punishments. Behn emphasizes the consequence of this shift as she proceeds to portray her narrator to encourage Oroonoko to obey the colonists or to prepare to become a victim of their violence: "He made me some Answers that shew'd a doubt in him, which made me ask him, what advantage it wou'd be to doubt? it would but give us a Fear of him, and possibly compel us to treat him so as I shou'd be very loath to behold: that is, it might occasion his Confinement."[119] This section of the scene is troubling. What had appeared to be a relationship of reciprocity and trust becomes one of inhumane dominance. As the "us" expands to include both women and men, Behn makes her narrator join in both the "Fear" of Oroonoko and the "loathsome treatment"—the enslavement—with which the colonists threaten him. Behn's narrator has moved her "racial alliances, from noble-ized female to colonizing white."[120]

Kaul characterizes the "drama of reciprocity" as "a process by which the representational strategy of anticonquest works itself out, and works toward establishing both the sentimental innocence *and* the cultural and material dominance of the imperial subject."[121] Though Kaul does not address this particular scene in Oroonoko, Doyle's reading certainly suggests this to be the case here. Indeed, she argues, "By the end of this scene her relation to Oroonoko has moved from one of perfect intimacy to one of stark dishonesty . . . . In the rest of the narrative, she serves as a spy while pretending to be a friend. In other words, she serves as the sentimental pivot by which common white becomes noble white becomes dominant white and noble black becomes common black becomes enslaved black."[122] Though I find Doyle's reading persuasive, I would suggest that recognizing the Epicurean basis of the narrator's relationship to Oroonoko helps us appreciate how the novel does not endorse but, rather, invites a critique of this progression. Doyle maintains that the narrator "serves as a spy," but there is no evidence that Behn's narrator informs male colonists of the private details she learns about Oroonoko through her practice of sensitive witnessing. Rather, Behn invites the reader to appreciate the awkwardness of her narrator's position as well as her incapacity, as a woman, to have a real influence on Oroonoko's fate.

Above, we recognized the "absurdity" of Oroonoko's character, but absurdity distinguishes the narrator's character as well. Oroonoko is a "royal slave"; Behn's narrator is an English woman. Both intersectional identities hybridize the dominant and subordinate. Behn encourages a recognition of the tragic consequence of her narrator's hybrid identity when she portrays her narrator explaining to Oroonoko that his behavior could "possibly compel us to treat him as I shou'd be very loath to behold." Taken on its own, this line could prompt a pessimistic reading that suggests it attempts to provide a rationalization for, and even defense of, abuse. However, situated within the work as whole—particularly, in relation to other passages in which the narrator is profoundly affected by Oroonoko's as well as Imoinda's suffering—the line instead appears to encapsulate the contradictory nature of the narrator's character. Having been born English, she has no choice but to identify with the colonists ("us"), but having been born a woman, she has no choice but to submit to (to be "compelled" by) the mechanisms of the colonial project and the slave trade on which it relies. This is therefore not an empty "it would hurt me more than it would hurt you" statement. The narrator's sensitivity to the suffering of others throughout the novella demands that we take her at her word when she says "I shou'd be very loath to behold" his anguish. Indeed, as Epicurean transcorporeality informs the narrator's practice of sensitive witnessing, it is as if Behn suggests that the narrator feels Oroonoko's suffering as her own. Admittedly, such sharing of feeling may be read as itself a form of imperialist appropriation and thus as an act of power on the narrator's part. However, it must also be acknowledged that the gender of Behn's narrator positions her to share feeling with Oroonoko, for neither of them is endowed with the power or autonomy that comes with having "property in his own person."[123] As a result, both are fundamentally vulnerable to harm.

Behn's sensitive witnessing reaches its highest pitch during the scenes of mutilation and torture that conclude the work. To analyze these scenes through the framework of Epicureanism is to discover the sublime nature of the knowledge-making that sensitive witnessing fosters. Though it would be a few more decades until Longinus's foundational treatise on the sublime, *Peri Hupsous*, would become "thoroughly current"[124] in Britain, the term appears in the final paragraph of *Oroonoko*: "Thus Dy'd this Great Man; worthy of a better Fate, and a more sublime Wit than mine to write his Praise; yet, I hope, the Reputation of my Pen is considerable enough to make his Glorious Name to survive to all Ages; with that of the Brave, the Beautiful, and the Constant *Imoinda*."[125] Though Behn here suggests that the work is *lacking* in sublimity, she nevertheless arouses the sublime with the profound emotion and arresting imagery she conveys throughout the work and especially in the final scenes in which her narrator witnesses Imoinda's and Oroonoko's gruesome deaths. Much as in her poem "To Creech," she distances herself from the sublime in a conventional performance of feminine humility and deference only then to proceed to cultivate it. Before describing Oroonoko's murder of Imoinda, the narrator portrays their final "parting"—their "eternal Leave taking"—as "very Moving."[126] Here, Behn evokes what would become a key feature of the eighteenth-century sublime: emotional "transport." The importance of this feature is illustrated in this passage from one of the earliest English translations of Longinus: "for the *Sublime* does not persuade, but creates Transport, always and every where superior to the *Persuasive* or the Elegant."[127] Behn portrays her narrator to be transported by the last farewell of Oroonoko and Imoinda, and such dramatic emotion continues until the novella's end.

Discomposure would become another key feature of the sublime, and this certainly characterizes both the narrator's and the reader's experience in the novella's denouement. To use the narrator's term, the images that Behn presents in the work's conclusion are "gashly."[128] According to William Smith's 1740 translation of Longinus, the sublime is "endued with Strength irresistible," a strength that "strikes home," "triumphs over," and thereby discomposes its witness.[129] As we have already observed, Behn's narrator feels compelled to "leave the Place" of Oroonoko's bodily mutilation "for some time."[130] Her inability to remain indicates the extent to which the scene has discomposed her and thus suggests its sublimity. Over half a century later, Edmund Burke would cite Milton's depiction of Death in *Paradise Lost* as an exemplary instance of the sublime (specifically, sublime "obscurity").[131] Behn too evokes the sublime with her portrayal of obscurity and death as her narrator bears witness to Oroonoko's awful transformation: "We ran all to see him; and, if before we thought him so beautiful a Sight, he was now so alter'd, that his Face was like a Death's Head black'd over; nothing but Teeth, and Eyeholes."[132] In this horrifying depiction, obscurity ("his Face . . . black'd over") gives way to the uncanny—Oroonoko's once familiar and "beautiful" visage emptied, metamorphosed into a darkened death mask. As it does so, the beautiful becomes the sublime, and the narrator is only just barely able to retain enough presence of mind to be able to continue to serve as a witness to the scene.

In his 1704 *Grounds of Criticism in Poetry*, John Dennis would depict the sublime as "an invincible force which commits a pleasing Rape upon the very Soul of the Reader."[133] Though Behn's narrator appears to experience more pain than pleasure at the horrifying sights that conclude *Oroonoko*, it is certainly the case that she feels overcome by, compelled to surrender to, them. In the Enlightenment era as now, the term *rape* signified "[t]he act of taking something by force"—"[o]riginally and chiefly: the act or crime, committed by a man, of forcing a woman to have sexual intercourse with him against her will, esp. by means of threats or violence." It could also denote "[t]he act of carrying off a person by force," which links it to the concept of "transport" discussed in relation to Longinus's *Peri Hupsous* above.[134] At the conclusion of *Oroonoko*, the narrator appears seized and even penetrated by Oroonoko. His story has gotten inside her—metaphorically, via the profound emotion she feels for him, but also literally as the "strong" scent of his mutilated body ("the earthly Smell") enters her body through her nostrils.[135] With the sublimity of his tale, as well as the atoms of his body, the "royal slave" crosses the threshold between self and other. Oroonoko is pierced by his own knife and then that of his executioner, while the narrator is pierced by what she witnesses. Both are, thus, ravished albeit with dramatically different results.

Indeed, Behn employs the terms *rape* and *ravish* at various points in the work. For instance, late in the work, while presenting his rationale for killing his beloved Imoinda, Oroonoko explains that he fears that, should she be allowed to survive, Imoinda would become "a Prey, or at best a Slave, to the inrag'd Multitude; his great Heart cou'd not indure that Thought. *Perhaps*, said he, *she may be first Ravished by every Brute; exposed first to their nasty Lusts, and then a shameful Death*."[136] Here, of course, the reference is to sexual violence. However, in an earlier passage, as the narrator describes the landscape of Surinam, she depicts "the Prospect" before her as "the most ravishing that fancy can create."[137] Here, with the term *ravishing*, Behn's narrator emphasizes the aesthetic sublimity of what she sees. Then, just a few paragraphs later, Behn uses the term very differently. Describing the body of a sheep that had been killed by a tiger, her narrator observes, "they found her [the tiger] muzzling in the Belly of a new ravish'd Sheep, which she had torn open."[138] In this case, which chillingly foreshadows Oroonoko's disembowelment and dismemberment that occur at the novel's conclusion, the term *ravish'd* signifies how the sheep's body had been penetrated violently, ravaged, and forcefully dragged away by the tiger prior to her feasting upon it. Though the term's meaning shifts to accommodate its particular context, the basic themes of penetration (whether pleasurable or violent) and being carried off by force (whether physically or emotionally) remain constant. The specter of rape haunts this work, and this haunting helps to accentuate the sublime nature of the knowledge-making that Behn's practice of sensitive witnessing entails.

Recognizing that the narrator of *Oroonoko* is "ravished" by what she witnesses, we must consider what distinguishes her from Doctor Baliardo in *The Emperor of the Moon* before we conclude our analysis of the novella. As we have seen, Baliardo is also "ravished" by his sense perceptions. However, whereas Baliardo's experience of ravishment leads him into error, Behn indicates that her narrator's capacity for being ravished is key to her effectiveness as a witness. What separates them is not their propensity to being ravished nor is it their sensitivity. In fact, together, these very different works insist that all humans—whether male or female, white or black, English or African—are sensitive creatures. Like the sheep being devoured by the tiger, they are sensitive beings simply by virtue of being material beings. Instead, their difference lies in how they accommodate their materiality and the sensitivity their materiality necessitates. Doctor Baliardo is laughably gullible because he denies both his materiality and his sensitivity, insisting on his "purity" and, thus, his masculine philosophical modesty. In stark contrast, *Oroonoko*'s narrator is highly perceptive and capable of unearthing the darkest and most private of secrets precisely because she recognizes her materiality and sensitivity, eschewing the purity of masculine modesty for the sake of a highly embodied feminine variety. Baliardo denies his materiality and sensitivity for the sake of his masculine empirical practice; *Oroonoko*'s narrator founds her empirical practice on her materiality and sensitivity.

The consequence of this difference is significant. Considered alongside one another, these works suggest that Baliardo is prone to being fooled—to being overwhelmed by, and thus losing his mind to, the "tickle" of sensation—because his faith in his own modesty and "purity" gives him a false sense of confidence and security. In contrast, the narrator of *Oroonoko* fully recognizes her own sensitivity, and doing so makes her realize that she must be vigilant so as not to be swept away by sense (and, therefore, potentially by error). Considered in this context, the narrator's decision "to leave the Place" of Oroonoko's mutilation "for some time" takes on new meaning. It attests to the narrator's recognition that, because of her sensitive nature, she must work to keep her presence of mind in an intentional, determined, and persistent fashion. By stepping away "for some time," she is capable of maintaining her status as a reliable witness as she proceeds to witness and to tell Oroonoko's story. Baliardo does not exactly "fall into Fits" as *Oroonoko*'s narrator imagines herself doing should she stay in the presence of such a "gashly sight" for too long, but the utter disconnection from empirical reality that he does experience is not all that different from a "fit." For Behn, there is no doubt that the feminine sensitive witness is preferable to the masculine modest witness. The sensitive witness is uniquely adept at discernment not only because she is able to gain special access, as we have seen Behn's narrator does with Oroonoko, but also because she knows she must take care not to lose her mind.

• • •

Behn's engagement with Epicurean materialism helps inspire her to recognize the seductiveness implicit within empirical observation. Material beings that we are, humans are prone to being "charmed" or "tickled" by what we perceive. To deny this propensity is to deny how easily it is for us to be swayed and, thus, fooled. Behn reveals that, in claiming purity and detachment from materiality, the modest witness positions himself as an easy target of deception. Believing himself to be aloof and neutral, he is likely to witness what is pleasurable while avoiding that which is disturbing or painful. He is also likely to claim that he has a complete and clear view. Behn insists that philosophical authority will only be legitimate if we abandon the dream of such modesty and recognize it for what it is: a fantastical and disabling fiction. As Markley argues, Behn expresses her "radical skepticism" about early modern philosophy by engaging in "ironic self-distancing."[139] We see such self-distancing, albeit in different forms, in the works that this chapter has discussed. However, there is an earnestness and even optimism in Behn's treatment of philosophy as well. This chapter has demonstrated that, though Behn is a skeptic, she is not a cynic when it comes to philosophical exploration.

By embracing the practice of sensitive witnessing she promotes, we may be "charmed" and "tickled" or we may be "sickened," but we will in the process become highly discerning observers, well-equipped to understand and to navigate the dizzying world of modernity. She invites us to base our knowledge on sublime encounters with the material world of which we are a part. Receptive to transport and discomposure, radically open to what the world presents, we become capable of registering the full power and variety of empirical details. Then and only then, Behn suggests, are we truly able to "survey all" and thus honor the truth of our human and extra-human material world. Suvir Kaul claims that "Behn's literary remembrance of Oroonoko and Imoinda, is . . . a memorial to the personal and cultural cruelties of the colonial plantation economy."[140] Behn's practice of sensitive witnessing is instrumental to this act of memorialization. The modest witness abstracts and, thus, obscures or obfuscates such cruelties; in so doing, he offers techniques to deny if not rationalize them. In contrast, Behn's sensitive witness demands that cruelties not only be seen but also felt. Though the reliability of her narrator's observations and the veracity of the story she tells may be uncertain, her approach to witnessing nevertheless calls for a reckoning. What the result of this reckoning may be is highly ambiguous and largely left to the reader to decide, but such ambiguity is itself an indication of Behn's critique of and refusal to conform to masculine modesty. Even as she makes Oroonoko "her slave" by writing his story, a deeply ironic (and perhaps satirical) form of "ownership," she ultimately resists allowing her work or her protagonist to be reified.[141] In contesting bodily integrity, Behn's Epicureanism presents a damning challenge to the property in persons (whether that property is in one's own person or in that of another). Feeling with and for Oroonoko's sublime suffering, the novella and its reader discover the irresistible power, as well as the ethico-philosophical value, of discomposure. For Behn, the best memorial is one that, like Oroonoko himself, refuses to be contained.

THREE

• • •

ESPIONAGE, SKEPTICISM, AND THE PROCESS OF PHILOSOPHY: ELIZA HAYWOOD

ELIZA HAYWOOD MAY BE best known for her amatory fiction, but her literary productions spanned many genres, topics, and styles. As various scholars have demonstrated, Haywood's works regularly engage with and even perform philosophical practice. In one such work, her periodical *The Female Spectator* (1744–46), Haywood develops and promotes a practice of sensitive witnessing that resembles the empirical practices pioneered by Cavendish, Hutchinson, and Behn explored in previous chapters. Like her predecessors, Haywood uses sensitive witnessing to assert the authority of female philosophers such as herself. Moreover, she too presents a critique of the ethics and efficacy of Baconian modest witnessing. However, her critique is distinctively amplified and pointed. I argue that, portrayed by Haywood, the sensitive witness becomes a voice for skepticism, and her portrayal contributes in significant ways to the Enlightenment tradition of "the skeptical sublime."[1]

Mary Beth Harris argues that, in her 1727 Oriental romance, *Philidore and Placentia; or, L'Amour trop Delicat*, Haywood poses a serious challenge to neoclassical masculinity as she offers "a pessimistic portrait of the predatory rational man" and links male homosocial desire to the fetishization of women.[2] This chapter builds on Harris's study by exploring how Haywood's *Female Spectator* pursues a corresponding critique of Baconian modesty. According to Harris, in *Philidore and Placentia*, Haywood "recalibrates the value of female passion and . . . brings neoclassical conceptions of rational virtue and male modesty under structural and cultural scrutiny."[3] I argue that Haywood performs a similar recalibration in *The Female Spectator*—often characterized as the first periodical in English written by a woman primarily for women—and that she challenges not only men's "predatory" sexual practices but also their philosophical practices that were widely thought to distinguish them from, and make them superior to, women. As Harris demonstrates, for Haywood, "[m]asculine virtue is not a counter-balance to female passion (i.e. rationality) as so many of Haywood's contemporaries claimed; rather, it is a philosophical manipulation designed to support men's cultural authority over women."[4] Like *Philidore and Placentia*, *The Female Spectator* reveals such manipulation as an act of "fabricat[ion]."[5] It insists that men and women are both, fundamentally, desiring beings.[6] Furthermore, it challenges the incompatibility of desire and reason. As Eve Tavor Bannet argues, "*The Female Spectator* vigorously contested *The Spectator's* attempts to exclude women from the public sphere . . . . [It] satirized, unmasked, and answered *The Spectator's* gender politics and put men of letters like her 'learned brother' in the serviceable and secondary place she designed for them in her woman-centered world."[7] Whereas Bannet shows how Haywood advocates for women's access to politics and education in her study, I focus primarily on how her periodical advocates for women's access to philosophy.

Haywood's interest in the dynamics and consequences of desire for philosophy links her work to that of Aphra Behn. Indeed, as Gabrielle Starr shows, the similarity of their interest in desire results in a formal similarity between their works. Both authors rely heavily on lyric conventions in their prose writing. Starr explains,

> As . . . Aphra Behn and Eliza Haywood sought ways to accomplish the affective requirements of their narratives, the substantial affinity between the province of the lyric and the demands of their writing led them to the use of lyric conventions. In *Love in Excess* (1719) Haywood reinterprets the metaphysical tradition of amatory poetry, especially that of Donne, to create her own figures of emotional excess. Behn works within the tradition of pastoral lyric as filtered through romance and her own poetry in *Love-Letters between a Nobleman and His Sister* (1684–87), using lyric to make the representation of emotion not merely descriptive, but affective as well.[8]

Starr goes on to assert, "These uses of lyric form a distinct counter-tradition to the concurrent wane of the lyric mode in Augustan poetry, moving from

Behn's contemporaries through the age of Pope."[9] Informed by Starr's analysis, I would contend that, for Haywood much as for Behn, it is the propensity of individuals, both female and male, to experience "emotional excess," as incited by the persistently penetrating quality of affective stimuli, that precludes the effectiveness and reliability of Baconian philosophical modesty. In this chapter, I argue that, like Behn, Haywood promotes sensitive witnessing as an alternative to Bacon's masculine modest variety; however, whereas Behn maintains a relatively optimistic view of sensitivity, as expressed by her Lucretian libertinism, Haywood's outlook commingles sensitivity with reason and is infused with skepticism.

Introducing *The Female Spectator*

As its title indicates, Haywood's periodical *The Female Spectator* capitalized on the popularity of Addison and Steele's *The Spectator*. Though she adopted or adapted many formal and stylistic characteristics of her periodical's male counterpart, the differences between the two publications are significant. With *The Female Spectator*, Haywood presents a satirical parody of Addison and Steele. This aspect of her periodical is immediately apparent as, in the first issue, she portrays its eponymous central character, the Female Spectator herself. Though her moniker closely corresponds to that of her "learned Brother of ever precious Memory," her personality and, even more importantly for our purposes, her empirical practice are strikingly distinct from those of her brother.[10] The latter depicts himself thus:

> I live in the World, rather as a Spectator of Mankind, than as one of the Species; by which means I have made my self a Speculative Statesman, Soldier, Merchant and Artizan, without ever medling with any Practical Part in Life. I am very well versed in the Theory of an Husband, or a Father, and can discern the Errors in the Oeconomy, Business, and Diversion of others, better than those who are engaged in them; as Standers-by discover Blots, which are apt to escape those who are in the Game. . . . In short, I have acted in all the parts of my Life as a Looker-on, which is the Character I intend to preserve in this Paper.[11]

As this introduction shows, Addison and Steele's Spectator prides himself on being removed—disengaged—from his environment. He suggests that because he is a "Looker-on," rather than one who meddles in the "Practical Part in Life," he has a high aptitude for discernment. His distance from society enables him to maintain what he calls "an exact Neutrality" and an "unprejudic'd Eye."[12] He even goes so far as to say that he has "all the advantages of a dumb man": "I have indulged my Silence to such an Extravagance, that the few who are intimate with me, answer my Smiles with concurrent Sentences, and argue to the very Point I shak'd my Head at without my speaking."[13] Here, the Spectator portrays his taciturnity as an indication of his characteristic detachment and, thus, of his authority as an observer. As this self-description shows, the Spectator prides himself on being a modest witness. As Ronald Paulson demonstrates, Mr. Spectator embodies a distinctive "form of disinterestedness" as he "modulates the austere virtue of civic humanism [advocated by Shaftesbury] into politeness."[14] Aloof and unmovable, the Spectator claims that his empirical practices and the social commentaries they inspire are incontrovertible because he does not get caught up in what he encounters. He is both Everyman and No-man; as such, he embodies Francis Bacon's ideal modest philosopher—he who inhabits the view from nowhere. That he hopes to "print [himself] out, if possible, before [he] die[s]" further emphasizes the abstract quality of his existence. It is almost as if he has no corporeal form.

Haywood envisions the central persona for her periodical much differently. On the one hand, there are important continuities between *The Spectator* and *The Female Spectator*. As David Marshall observes, "critics increasingly emphasized the subjective response and experience of readers and beholders."[15] This was certainly the case in both periodicals. Marshall observes, "Addison . . . drew upon Locke to inaugurate a psychological or empirical aesthetics" grounded in personal experience. The same is true for Haywood in *The Female Spectator*. However, whereas the Spectator derives his authority from his detachment from everyday affairs, it is the Female Spectator's very embeddedness in, and embodied experience of, society that informs her and lends credence to her observations. In the periodical's first issue, she explains,

> I have run through as many Scenes of Vanity and Folly as the greatest Coquet of them all.—Dress, Equipage, and Flattery, were the Idols of my Heart. I should have thought that Day lost which did not present me with some new Opportunity of shewing myself.—My Life, for some Years, was a continued Round of what I then called Pleasure, and my whole Time engross'd by a Hurry of promiscuous Diversions.—But whatever Inconveniences such a manner of Conduct has brought upon myself, I have this Consolation, to think that the Publick may reap some Benefit from it.[16]

According to Haywood's portrayal, the Female Spectator is particularly well suited to bringing "Benefit" to "the Publick" because of her wealth of experience. Having spent several years as a coquette and living a life that "was a continued Round of what [she] then called Pleasure," the Female Spectator is well versed in the ways of the world. Though she acknowledges that her former promiscuousness has had negative consequences ("Inconveniences") in her personal life, she asserts that it also affords her significant authority as a witness, writer, and teacher. Whereas the Spectator's authority derives from "disengagement" from society, it is the Female Spectator's personal history of engagement in society from which she derives her knowledge, understanding, and thus expertise. Characterized by Paulson as a "civic humanist gentleman," Mr. Spectator claims that it is because he has never been "in the game" of society that he is capable of superior "discernment."[17] In direct contrast, the Female Spectator asserts that it is precisely because she has lived a long time "engrossed" in the game of society that she is well prepared to analyze society and to offer lessons to her readers that are informed by this analysis. Though the Female Spectator shares her brother's interest in everyday life and how individuals encounter and perceive it, she is *absorbed* by her material environment while he lives at a remove from his.[18] The Spectator's knowledge is "speculative" and based on "theory"; the Female Spectator's knowledge is profoundly engaged, embodied, and experiential.

Like her brother, the Female Spectator employs collaborators to assist her in her observations, but their different portrayals of their collaborators also highlight the differences in their approaches. In his first paper, the Spectator explains that his periodical is organized around "a Plan" that has been "laid and concerted . . . in a Club," which "meets . . . on *Tuesdays* and *Thursdays*," and is supported by "a Committee" that "sit[s] every Night, for the Inspection of all Papers"—that is, all letters from the periodical's readers.[19] In the second paper, he proceeds to introduce the six members of his club: Roger de Coverly, "a Baronet" who is "of antient Descent"; an unnamed "Batchelour, who is a Member of the *Inner*-Temple" and "has chosen his Place of Residence rather to obey the Direction of an old humoursome Father than in Pursuit of his own Inclinations"; Sir Andrew Freeport, "a Merchant of great Eminence in the City of *London*"; Captain Sentry, "a Gentleman of great Courage, good Understanding, but invincible Modesty"; Will Honeycomb, "a Gentleman" who has "ever been very careful of his Person, and always had a very easy Fortune"; and an unnamed "Clergyman, a very philosophick Man, of general Learning, great Sanctity of Life, and the most exact Good Breeding."[20] The Spectator explains that it is because "all Matters of Importance" are "concerted . . . in a Club" that he has chosen to develop his periodical with the assistance of several collaborators.[21]

Though the Female Spectator similarly relies on collaborators, the rationale she provides for doing so is different from her brother's and, I would argue, offers another testament to the value she places on embodied and engaged experience. Offering a conventional expression of feminine humility and deference, she explains that, after "setting down many Things, which, being pleasing to myself, I imagin'd would be so to others; but on examining them the next Day, I found an infinite Deficiency both in Matter and Stile" so that "there was an absolute Necessity for me to call in to my Assistance such of my Acquaintance as were qualified for that Purpose."[22] She proceeds to introduce three collaborators. The first, whom she "distinguish[es] by the Name of *Mira*" is a married woman characterized by "sparkling Ideas, . . . receiv'd from Nature and Education."[23] The second she simply designates as "a Widow of Quality" and describes as characterized by "Vivacity" and as a "Confidante of Secrets" to "those of her Acquaintance."[24] The third is "the Daughter of a wealthy Merchant" and is a "fine young Creature" whom the Female Spectator "call[s] *Euphrosine*, since she has all the Chearfulness and Sweetness ascribed to that Goddess."[25] The Female Spectator goes on to explain, "whatever Productions I shall be favour'd with from these Ladies, or any others I may hereafter correspond with, will be exhibited under the general Title of *The Female Spectator*, and how many Contributors soever there may happen to be to the work, they are to be consider'd only as several Members of one Body, of which I am the Mouth." The Female Spectator's portrayal of her relationship to her collaborators is significantly different than that of her brother. Indicating that he and his collaborators hold regular meetings ("on *Tuesdays* and *Thursdays*") as a "club," the latter associates himself and his "Gentleman" collaborators with contemporary coffee house culture and grants his group a quasi-formal, or at least well-organized, status within the public sphere.[26] As the Spectator's epigraph to Paper 2 from Juvenal suggests, they are united in "voice": "*Ast Alii sex / Et plures uno conclamant ore.*"[27] In contrast, the Female Spectator and her collaborators are joined as if into one corporeal form ("one body").[28] Though the Spectator presents his collaborators as real people, his portrayal of their relationship nevertheless evokes the aloofness, detachment, disembodiment, and consequent abstraction associated with masculine modesty, whereas the

Female Spectator emphasizes her and her collaborators' shared physicality. Such physicality is further reinforced by the frontispiece appended to the 1746 collected edition of *The Female Spectator*, which portrays the Female Spectators sitting so close to one another that they are almost touching (see figure 1).[29]

Another significant difference between the Female Spectator's introduction to her periodical and her brother's is that whereas the latter merely seeks assistance from "a committee" of collaborators, the former also employs a collection of "spies." The Spectator explains, "[w]e have appointed a Committee to sit every Night, for the Inspection of all . . . Papers as may contribute to the Advancement of the Publick Weal."[30] Mr. Spectator invites any readers "who have a mind to correspond with [him]" to submit letters so that he, together with his fellow committee members, might reflect on them and consider whether or not they are valuable enough to the "Publick Weal" to deserve to be incorporated into his periodical.[31] From the outset, then, Addison depicts *The Spectator* as a fundamentally collaborative production in which he and his fellow committee members, together with their readers, form a public-minded participatory collective. According to the Spectator's depiction, it is from this broad collective that the stories and perspectives of the periodical will be drawn. He gives the impression that he has a plentitude of knowledge, not only due to his openness to his readers' contributions or the variousness of his committee members' biographies but also to his own wide-ranging observations: "when I consider how much I have seen, read and heard, I begin to blame my own Taciturnity; and since I have neither Time nor Inclination to communicate the Fulness of my Heart in Speech, I am resolved to do it in Writing."[32] As this passage illustrates, Mr. Spectator believes himself to be an abundant repository of knowledge and experience as a result of all that he has, thanks to his distinctive reservedness or modesty (his "taciturnity"), "seen, read, and heard."

Whereas Mr. Spectator portrays his view to extend far and wide, with only the help of his small committee of friends and his readers' periodic correspondence, the Female Spectator feels compelled to seek additional help in extending her view. She explains:

> to secure an eternal Fund of Intelligence, Spies are placed not only in all the Places of Resort in and about this great Metropolis, but at *Bath*, *Tunbridge*, and the *Spaw*; and Means found out to extend my Speculations even as far as *France*, *Rome*, *Germany*, and other foreign Parts, so that nothing curious or worthy of Remark can escape me; and this I look

FIGURE 1: Eliza Fowler Haywood, *The Female Spectator* (London, 1744). Courtesy of the Beinecke Rare Book and Manuscript Library, Yale University.

> upon to be a more effectual Way of penetrating into the Mysteries of the Alcove, the Cabinet, or Field, than if I had the Power of Invisibility, or could with a Wish transport myself wherever I pleased, since with the Aid of those supernatural Gifts, I could still be in no more than one Place at a Time; whereas now, by tumbling over a few Papers from my Emissaries, I have all the Secrets of *Europe*, at least such of them as are proper for my Purpose, laid open at one View.[33]

As we have seen, like Mr. Spectator, the Female Spectator relies on a group of collaborators. She also similarly invites her readers to correspond with her and, thus, to become "Contributors" to the periodical.[34] Indeed, she goes so far as to designate her reader-correspondents as "Members" of the Female Spectator's "Body" along with her three collaborators.[35] However, she nevertheless indicates that she requires even further assistance to be able to fulfill her goals. Having just acknowledged her employment of spies, she proceeds to defend herself from the accusation that she is motivated by the "Design of gratifying a vicious Propensity of propagating Scandal": "Whoever sits down to read me with this View, will find themselves mistaken."[36] Her goal is to "bring real Facts on the Stage," but she is determined to "conceal the Actors Names under such as will be conformable to their Characters."[37] She aims "to expose the Vice, not the Person."[38] As this comment makes clear, though, the revelation of hidden "Facts" and secret, or at least private, stories ("exposure") is her main intent. She may not seek to call out specific individuals, but she does intend to call out their behaviors (specifically, their "vices") and, in doing so, to illuminate what is generally obscured from view. Mr. Spectator's goals seem humble by comparison. He may have double the number of committee members as the Female Spectator has. However, the former appears content with what he, his committee members, and his readers have "seen, read and heard," whereas the latter is not. She refuses to be satisfied until she has "all the Secrets of *Europe*, at least such of them as are proper for [her] Purpose, laid open at one View," and to satisfy this ambition she must rely on an army of witnesses.[39]

That the Female Spectator employs a team of spies whereas Mr. Spectator does not suggests not only a quantitative but also a qualitative difference in their empirical methods. As a commander of spies, as well as a leader of a committee and a correspondent with a wide readership, the Female Spectator has more witnesses at her disposal than her male counterpart does. Even more importantly, though, is the perspective that she suggests her spies afford her. As she explains in the quote provided above, she aims for an intimate encounter with all that is "curious or worthy of Remark."[40] With her spies, she is able to "penetrate . . . into the Mysteries of the Alcove, the Cabinet, or Field."[41] Mr. Spectator is proud of his ability to conduct his observations as a "detached" "looker-on" and, therefore, at a perspectival distance from what he witnesses. In contrast, the Female Spectator prides herself on her ability to conduct her observations at close range with the help of her secret agents. Her spies manage to gain access to places where observers aren't typically allowed to be. Consequently, Haywood suggests that the Female Spectator's empirical aptitude surpasses that of her male counterpart in its extensiveness. Mr. Spectator's observations may be broad, but they are not as deep as the Female Spectator's.

I would argue that it is precisely because the Female Spectator eschews the masculine modesty on which Mr. Spectator prides himself that she is inclined and able to cultivate such a "penetrating" practice of exploration. As we have observed in previous chapters, such masculine modesty and the empirical detachment that it entails were generally not perceived to be available to women because of their presumed association with, and consequent inability to transcend, matter. As Paulson observes, it is significant that "Addison's charmed circle" is "a male group" and "Mr. Spectator . . . [a] civic humanist gentleman."[42] Instead of resigning herself to philosophical inferiority, though, Haywood like Behn, Cavendish, and Hutchinson before her transforms the perceived liability of her gender into an advantage. She cedes the privilege of neutrality to her male counterpart but claims the privilege of *access* for herself. Associated with matter as well as the private domain because of her gender, the Female Spectator is able to "penetrate" and thus perceive what is inaccessible to her brother.

This is precisely what she proceeds to do. Issue after issue is devoted to highly personal, private, and often secret and even scandalous stories. Her claim that her spies endow her with quasi-"supernatural" powers ("of Invisibility" and teleportation) seems fully supported by how fully embedded the Female Spectator is in the private lives of strangers. That she often withholds precise details regarding how she has gained access to the stories she tells only reinforces the reader's sense of her powers and her profound "engrossment" in the material world. Take, for example, the story of Belinda, a "vain, wild, unthinking Creature," who proceeds through a series of illicit and secretive amours that persist even after she has been married and become a Countess.[43]

The Female Spectator neglects to specify exactly how she has become privy to this episodic and multilayered story, which includes multiple private conversations and furtive infidelities. She only once has recourse to the phrase, "[r]umour will have it."[44] Otherwise, it would seem that the details she provides are based on direct empirical observation, given their specificity and the confidence with which she presents them. For instance, when Adonius sets his sights on Belinda, his wife—"the young, the beautiful, the tender Amadea"—"pines in secret, without daring to complain, and now too late regrets her easy Faith, which flatter'd her with the Hopes of securing to herself so mutable a Heart."[45] So profoundly engaged is the Female Spectator in the world that she is capable of accessing the most secret of facts and feelings. The most private and even internal experiences of others are not beyond her reach.

The Female Spectator's Sympathy: Feeling For and With

Thus does the Female Spectator "penetrate . . . into the Mysteries of the Alcove, the Cabinet, or Field," and such penetration is not a one-way street.[46] She regularly allows herself to be emotionally affected by what she observes. For instance, during her narration of Belinda's story, she exclaims, "What sad Effects do giving way to any Passion, tho' of the most tender Kind, produce, especially in our Sex! . . . What Woe, what Misery, what Despair wou'd then be the Lot of so every-way an abandon'd Wife!"[47] As these exclamations demonstrate, the Female Spectator does not witness the stories she tells at an emotional distance as her aloof brother does. Rather, she opens herself up to feeling that, as is the case in this passage, seems to swell beyond her discursive expression. She points to but does not detail the "sad Effects" and the "Woe, . . . Misery" and "Despair" that arise from "giving way to . . . Passion." It is as if they are so profound as to be ineffable. Moreover, the Female Spectator appears to be herself touched by Belinda's experiences and the emotions they incite. Indeed, she seems to take Belinda's experiences and emotions as her own while encouraging her readers to do so as well.

The Female Spectator's impulse to adopt characters' feelings as her own is closely associated with her pedagogical method, which is based on a fundamental identification between subject, teacher, and student. As was typical of the eighteenth-century periodical, and especially of those intended for a female audience, *The Female Spectator* was designed not only to entertain but also to educate its readers. As Jennie Batchelor and Manushag N. Powell explain in a discussion of *The Female Spectator*, "Periodicals . . . were ever so much more than repositories of news and opinion. They could be, to use a term also employed by Haywood, 'visionary,' moving their readers toward a more satisfying social structure using an educational programme of heterogeneous reading."[48] Moreover, the method of instruction that Haywood employs in *The Female Spectator* is one of precept by example. In her introduction to the periodical, the Female Spectator explains, "From my Observations of human Nature, I found that Curiosity had, more or less, a Share in every Breast; and my Business, therefore, was to hit this reigning Humour in such a Manner, as that the Gratification it should receive from being made acquainted with other People's Affairs, should at the same Time teach every one to regulate their own."[49] Sympathy is fundamental to Haywood's pedagogical approach. She seeks to encourage her readers to translate and thereby apply others' experiences ("other People's Affairs") to their own. Such translation requires readers to be able to feel both for and with the individuals that Haywood depicts.

I would suggest that the variety of sympathy that Haywood portrays and encourages in *The Female Spectator* is akin to that theorized by David Hume in his *Treatise of Human Nature*. Like Hume's, Haywood's variety of sympathy "depends upon the general resemblance between people" and "is predicated on our perception of a resemblance between another person and our idea of ourselves. . . . The 'force' or 'vivacity' that causes the conversion of sympathy comes from some aspect of the mind perceiving resemblance."[50] As Hume explains, "[R]esemblance converts the idea into an impression, not only by means of the relation, and by transfusing the original vivacity into the related idea; but also by presenting such materials as take fire from the least spark."[51] Like Hume, Haywood envisions a "conversion" of the stories she tells into "impressions" experienced by her readers. Moreover, it is because she presumes a fundamental "resemblance" between individuals, much as Hume does, that she is able to imagine such conversion transpiring smoothly, readily, and deeply.[52] As the Female Spectator "penetrates . . . into the Mysteries of the Alcove, the Cabinet, or Field," she allows herself and her readers to be penetrated by the characters and stories her expeditions uncover.[53] It is as they are thus penetrated that they become capable of learning from others' experiences so that they may "regulate" their own.[54] Of Haywood's amatory novel *Love in Excess*, Stephen Ahern argues, "The ability of lovers to connect in Haywood's novel hinges on the notion of sympathy, or more precisely on the notion of compassion in the original sense of the word, that is, the capacity not simply to pity another person's sufferings from a distance, but to suffer together with another, to participate in suffering."[55] I would suggest that Haywood's *Female Spectator* hinges on such compassion as well. Moreover, with the latter work she shows that people need not be lovers to experience it. In fact, even strangers may do so, as in the relationship she envisions between her readers and her characters.

Addison and Steele's *The Spectator* also relies on and seeks to stimulate sympathy; however, the variety of sympathy that it entails is significantly different from that entailed by Haywood's *The Female Spectator.* Their different versions of sympathy inform, and directly contribute to the differences between, the methods of witnessing that they practice and encourage. Whereas the sympathy that Haywood portrays in *The Female Spectator* resembles that theorized by Hume, the sympathy that Addison and Steele model and promote in *The Spectator* is more closely aligned with that theorized by Adam Smith. As Jonathan Lamb explains,

> Smith seeks to find in economic as well as affective life some form of restraint exercised through the medium of an invisible presence that sets passion the task of moderating itself. Thus the invisible hand of the market transforms the desire for idleness into technological improvement and productive expansion; meanwhile the impartial observer teaches us to moderate impatience by making us sensitive to shame, to restrain ambition by inclining us to sympathy, and so on.[56]

In his famous formulation, Smith argues that when we experience sympathy,

> we place ourselves in [the sufferer's] situation, we conceive ourselves enduring all the same torments, we enter as it were into his body, and become in some measure the same person with him, and thence form some idea of his sensations, and even feel something which, though weaker in degree, is not altogether unlike them. His agonies, when they are thus brought home to ourselves, when we have thus adopted and made them our own, begin at last to affect us, and then we tremble and shudder at the thought of what he feels.[57]

For Smith, sympathy is "founded" on an "imaginary change of situation." However, because "[t]he thought of their own safety, the thought that they themselves are not really the sufferers, continually intrudes itself upon them," it requires individuals to become impartial spectators.[58] Smith explains, it is only by "reduc[ing] the violence of the passions" to a "pitch of moderation" that one becomes capable of "entering into" scenarios that trigger powerful feelings—"situations which . . . bear hard . . . upon human nature."[59] Whereas for Hume and Haywood, individuals are readily able to enter into and thus experience another's suffering because of their fundamental resemblance, for Smith resemblance is not enough to enable sympathy because of how preoccupied we are with our own personal situation.[60] To overcome such preoccupation, according to Smith, we must experience the sufferer's feelings (and our own) with impartiality.

Though *The Spectator* precedes Smith's *Theory of Moral Sentiments* by nearly fifty years, the distinctive variety of sympathy that Addison and Steele model and promote with *The Spectator* anticipates Smith's. Fred Parker even goes so far as to characterize Smith's *Theory* as "an extrapolation of the principles underlying *The Spectator*."[61] Impartiality and detachment are the primary sources of Mr. Spectator's authority for they allow readers to count on him not only to judge rightly but also to sympathize readily with a wide array of experiences. Take, for instance, this opening from a correspondent's letter, printed in no. 336:

> There is an Evil which has prevailed from Generation to Generation, which grey Hairs and tyrannical Custom continue to support; I hope your spectatorial Authority will give a seasonable Check to the Spread of the Infection; I mean old Men's overbearing the strongest Sense of their Juniors by the meer Force of Seniority; so that for a young Man in the Bloom of Life and Vigour of Age to give a reasonable Contradiction to his Elders, is esteemed an unpardonable Insolence, and regarded as a reversing the Decrees of Nature.[62]

As this passage shows, this correspondent presumes that Mr. Spectator will be able to sympathize with the frustration that he experiences with his elders when they expect him to acquiesce to their point of view simply because they are older than he is ("the meer Force of Seniority"). Though this reader recognizes that the Spectator is not himself a youth, because of his confidence in Mr. Spectator's impartiality and consequent facility for sympathy, he nevertheless asks him to "assert the Rights of Youth" and thereby "prevent the pernicious Incroachments of Age."[63] Mr. Spectator offers no response to or commentary on the youth's letter and thus neglects to indicate explicitly what his reaction is to its sentiments, including whether or not he agrees with them. However, that he prints the youth's letter without an accompanying comment is itself an expression of his impartiality. So neutral and detached is he that, in this paper as in many others like it, it is as if Mr. Spectator drops out of the periodical altogether, allowing the points his correspondents make and the stories he tells to speak for themselves without any interference from his perspective.[64] However, at the same time, by printing the youth's letter and letting it speak for itself, he implies his assent as well as his sympathy with the youth. Moreover, he implicitly invites his readers to emulate his impartiality—his detachment and modesty—so that they too may sympathize with the youth, whether they themselves are young or old, and thus help foster the social change the youth seeks. As an embodiment of what Smith would decades later portray as the "impartial spectator," Mr. Spectator trains his readers to detach themselves from their own experiences as well as from those of others so that they may use their imaginations to "become in some measure the same person with" the youth as well as with the many other figures with whom he populates his papers. Whereas Haywood's Female Spectator is a sensitive witness who deeply sympathizes with her subjects by allowing their experiences and emotions to penetrate her so that she might herself feel them as and like they do, Mr. Spectator is a modest witness who sympathizes with his subjects by *imagining* at a distance what it is like to be them.

I would argue that such differences are reflective of the divergent goals of the two periodicals. Though both Addison and Haywood generally strive to "contribute to the Advancement of the Publick Weal," their intended means of so contributing are distinct.[65] As is evident from the periodical's opening paragraphs, Haywood's primary aim in *The Female Spectator* is pedagogical: specifically, as we have already seen, she aims to instruct readers how to "regulate" their own individual "Affairs."[66] In contrast, *The Spectator* focuses mainly on pleasure rather than on learning. Consider, for instance, the epigraph from Horace with which its first paper begins: "Not smoke after flame does he plan to give, but after smoke the light, that then he may set forth striking and wondrous tales."[67] This paired with the opening sentences of the first paper, which envisions the reader "perus[ing]" the publication "with Pleasure," sets an important foundation for the periodical as a whole.[68] From the outset, the reader's pleasure (and secondarily their "Curiosity, which is so natural to a Reader") occupy the Spectator's attention.[69] Later in this inaugural paper, he outlines his vision for the periodical thus: "I shall publish a Sheet-full of Thoughts every Morning, for the Benefit of my Contemporaries; and if I can any way contribute to the Diversion or Improvement of the Country in which I live, I shall leave it, when I am summoned out of it, with the secret Satisfaction of thinking that I have not Lived in vain."[70] The Spectator here reinforces his earlier appeal to readers' pleasure as he expresses his intent to provide "diversion." Though, according to this passage," "Improvement of the Country" is another primary goal of the periodical, in contrast to his sister, the Spectator offers little explanation regarding how such improvement is to take place.

Furthermore, whereas his sister proceeds to make a habit of gleaning lessons from the anecdotes she shares and the letters she includes, and of making these lessons explicit to her readers, the Spectator often narrates anecdotes without explicitly highlighting their lessons. Indeed, many of his papers lack even an implicit lesson. Though he may aim to "recover" his readers "out of that desperate State of Vice and Folly into which the Age is fallen" by "enliven[ing] Morality with Wit" and "temper[ing] Wit with Morality," the moral messages of the periodical are regularly obscure and not nearly as obvious as those in *The Female Spectator*.[71] A good case in point is the most famous of the Spectator's papers—nos. 411–421 on "The Pleasures of the Imagination." As the concluding sentence of no. 411 illustrates, these papers not only concern themselves with pleasure. They also stimulate pleasure as they provoke the reader's thought, satisfy their curiosity, and help them find new ways of pursuing pleasure: "I have in this Paper, by way of Introduction, settled the Notion of those Pleasures of the Imagination, which are the Subject of my present Undertaking, and endeavoured, by several Considerations, to recommend to my Reader the Pursuit of those Pleasures."[72] Whereas Haywood's main goal is to teach her readers how to achieve greater control over their lives—a goal that links her periodical to the conduct-book tradition—Addison and Steele's is to stimulate and help satisfy their readers' capacity for enjoyment by activating their minds.[73]

The modest witnessing that Addison and Steele model and encourage is conducive to their goal of innocent diversion, while the practice of sensitive witnessing that Haywood devises is conducive to her positive pedagogical goal. The Spectator serves as a guide for the reader, leading them through a broad variety of sites within English, and particularly London, society and keeping them healthfully occupied. "The Mind that lies fallow but a single Day," the Spectator observes, "sprouts up in Follies that are only to be killed by a constant and assiduous Culture."[74] To prevent such "sprouting" of "Follies," he aims to bring "Philosophy out of Closets and Libraries, Schools and Colleges, to dwell in Clubs and Assemblies, at Tea-Tables, and in Coffee-Houses."[75] In other words, the Spectator collects and curates human behavior and knowledge ("Philosophy") for his readers and offers it to them in their private gathering places ("Clubs," "Assemblies," "Tea-Tables," "Coffee-Houses"), allowing them to find ways to occupy their minds without having to leave their familiar environments.[76] Such occupation is able to remain wholesome, though, because the pleasure it affords is experienced modestly and, thus, at a distance. It is therefore a muted pleasure, perhaps better characterized as a detached sense of satisfaction. Soberly experienced, then, their pleasure only goes so deep. In reading *The Spectator*, the reader becomes like the Spectator himself—"a Looker-on" who is "well versed in the Theory of an Husband, or a Father, and can discern the Errors in the Oeconomy, Business, and Diversion of others, better than those who are engaged in them."[77]

Viewing human behavior and knowledge at both a physical and psychological distance, they become discerning observers, capable of "discover[ing] Blots, which are apt to escape those who are in the Game."[78] Kathleen Lubey argues that, in the Pleasures of the Imagination papers, "Addison's explanation of description's effects locates their force in the depth with which they strike the subject: so private, so interior is the pleasure taken in description that its most immediate effects leave their mark on the blood, or the humours, or the passions."[79] However, though Lubey extends such literary palpability to the entirety of *The Spectator*, I would argue instead that, throughout much of the rest of *The Spectator*, Addison and Steele counteract such "depth" of literary experience by encouraging their reader to be like the Spectator himself—a "looker-on" who cultivates neutrality and discernment by *not* becoming "marked" by what they experience (in literature or otherwise). The Spectator observes, "The Fancy must be warm, to retain the Print of those Images it hath received from outward Objects."[80] Taken as a whole, though, *The Spectator* registers ambivalence about the capacity of literature (and, for that matter, life) to "warm" the reader.[81] In fact, more flâneur than man of feeling, his example shows how to resist such warmth in order to maintain a reliable capacity for discernment.[82] If the Spectator has a "warm Fancy," it is so suppressed as to be undetectable.

Haywood also hopes to produce discerning observers, but the discernment she seeks to cultivate is based on deep sympathy and a sense of personal identification rather than, as is the case with Addison and Steele, affective remove. Whereas the Spectator encourages the reader to be a "looker-on"—a cultural tourist who observes without participating, even virtually, in what they witness—his sister aims to engage the reader in such a personal and emotional way that they become capable of experiencing the stories she depicts as if they are the reader's own. A recurring theme throughout *The Female Spectator* is how women's lack of experience in the ways of the world readily makes them naïve, gullible, and therefore vulnerable to victimization. With the periodical, Haywood seeks to counteract this contemporary trend by providing readers (specifically, female readers) with virtual experiences that will serve as stand-ins for actual ones. Haywood thus develops a practice of "virtual witnessing" that resembles, though in its overt sensitivity is distinguished from, that developed and practiced by members of the Royal Society.[83] In an analysis of Haywood's amatory novel, *Lasselia: or, the Self-Abandoned* (1723), Lubey asserts that, in reading Haywood's fiction, "readers come to acquire the most essential knowledge regarding the workings of human consciousness and desire, a knowledge they will convert into active self-scrutiny and self-government in social and sexual realms."[84] Lubey even goes so far as to argue that "[b]y granting readers virtual access to the desires and seductions of beautiful, psychologically sophisticated women characters, Haywood invents a specific type of eroticism—an amatory aesthetic," which "creat[ed] the imaginative conditions that were thought to please and instruct in the literary culture of the period."[85] I would argue that this "amatory aesthetic" is at play in *The Female Spectator* much as it is in Haywood's novels. Indeed, I would contend that, in *The Female Spectator*, Haywood uses this aesthetic to help mold not only polite and resilient women but also assertive and accomplished female philosophers.

Consider, for instance, one of the first characters Haywood includes within *The Female Spectator*: "the beautiful *Martesia*."[86] This young woman is "descended of the most illustrious Race, possess'd of all that Dignity of Sentiment befitting her high Birth, endued by Nature with a surprising Wit, Judgment, and Penetration, and improved by every Aid of Education."[87] However, in spite of such advantages, she is an easy target for predatory men, and Haywood attributes her vulnerability to her lack of experience. Because of "[h]er unexperienced Heart," she enters into a marriage in which "soon, too soon the fleeting Pleasures fled, and in their Room Anguish and Bitterness of Heart succeeded."[88] She then proceeds to become easily charmed by "the young *Clitander*," who "was just return'd from his Travels, had a handsome Person, an infinity of Gaiety, and a certain Something in his Air and Deportment which had been destructive to the Peace and Reputation of many of our Sex."[89] "Tho' no Woman could have more refin'd and delicate Notions than

Martesia," the Female Spectator observes, "all were ineffectual against the Sollicitations of her adored *Clitander.*—One fatal Moment destroy'd at once all her own exalted Ideas of Honour and Reputation, and the Principles early instill'd into her Mind by her virtuous Preceptors."[90] In the end, she "find[s] herself deserted by her Relations, and the greatest Part of her Acquaintance, without Love, without Respect, and reduced to the Pity of those, who, perhaps, had nothing but a greater Share of Circumspection to boast of."[91] After having given birth to a daughter—conceived during her affair with Clitander—"who expir'd almost as soon as born," Martesia banishes herself from England and ultimately ends up "roaming round the World in fruitless Search of that Tranquility she could not have fail'd enjoying at home in the Bosom of a Consort equally belov'd as loving."[92]

If only England were more like France, Haywood wishes, where "the People are, questionless, the gayest and most alert in the World, and allow the greatest Liberties to their Women."[93] According to Haywood, "to hear of a clandestine Marriage among [the French] is a kind of Prodigy, and tho' no Place affords Scenes of Gallantry equal to it in any Degree of Proportion, yet I believe there is none where fewer false Steps are made, or Husbands have less Reason to complain of the want of Chastity in their Wives."[94] Haywood here suggests that French women are less vulnerable to temptation and victimization because they are, in their "liberty," generally afforded more opportunities to gain, and thus learn from, worldly experiences than English women are. A main goal of Haywood's as an author is to remedy the lack of experience from which she believes many English women to suffer. She pursues this goal in her novels, as Lubey has shown, as well as in her periodical *The Female Spectator.* In doing so, she seeks to resist, and thus counteract the effect of, the trend of misrepresenting "Passion" that Haywood argues is common amongst contemporary "Romances, Novels, and Plays" wherein "the Authors seem to lay out all their Art in rendering that Character most interesting, which most sets at Defiance all the Obligations, by the strict Observance of which Love can alone become a Virtue."[95] According to Haywood, "such authors dress their *Cupid* up in Roses, call him the God of soft Desires, and ever-springing Joys, yet at the same Time give him the vindictive Fury, and the Rage of *Mars*—Shew him impatient of Controul, and trampling over all the Ties of Duty, Friendship, or natural Affection, yet make the Motive sanctify the Crime." She proceeds to observe, "How fatal, how pernicious to a young and unexperienced Mind must be such Maxims, especially when dress'd up in all the Pomp of Words!"[96] In contrast to the authors of "such Maxims," Haywood seeks to tell it like it is while ranging broadly throughout English society. She seeks to extend "liberties" to her English readers akin to those that French women enjoy, albeit experienced virtually through reading, so that they may no longer suffer the often dramatic consequences of the naivety and ignorance that a sheltered life produces.

The sensitive witnessing that Haywood models and promotes in *The Female Spectator* is instrumental in enabling not only the reader's virtual experience of such liberty but also their ability to learn from the experience and to translate it into personal lessons that they may apply to their own lives. As we have already recognized, it allows them to sympathize deeply and personally with the characters that Haywood depicts in the periodical. However, by allowing them to feel for and with characters, it also allows them to live vicariously through them. Thus, through sensitive witnessing, Haywood's readers are able themselves to experience the variable emotional states of Haywood's characters and, as they do so, they experience the challenges, choices, and consequences (both rewards and punishments) her characters face as if they were their own. Haywood thereby allows her readers to learn from her characters' successes and mistakes with the hope that, by encouraging them to be sensitive to the consequences of her characters' choices, her readers will be able to avoid making the same mistakes that her characters do in their own lives. To do so, they must learn how to resist dangerous temptation and to make good choices. By immediately confronting readers, whom she has trained to be sensitive, with the consequences of her characters' choices, Haywood seeks to prompt readers to become motivated and empowered to make their own choices with their own best interests in mind. This is what Haywood means when she expresses her intention to encourage readers to learn how to "regulate their own [affairs]."[97] Haywood may at times seek to cultivate emotional excess as she trains her readers in the practice of sensitive witnessing. However, her ultimate goal is to help her readers learn how to control and moderate such excess.[98]

The practice of sensitive witnessing, and the deep sympathy that enables it, make her readers receptive to such learning. Consider, for instance, the story of Amasina and Palamon. Amasina's

> high Birth, and the Accomplishments she was Mistress of heighten'd the Graces of her Person, and scarce any Age ever produced an Object of more universal Adoration. But of all the Addresses made to her, those of

> *Palamon* were the most countenanc'd by her noble parents, and agreeable to herself:—His Virtue, good Sense, and Breeding, made him respected by them, as the Gracefulness of his Person gave him the Advantage in her Eyes, above all others who pretended to her, tho' some there were whose Estates were far superior, and whose Declarations of Love were also accompanied with a greater Shew of Vehemence.[99]

In spite of her affection for Palamon, however, Amasina proceeds to become "a perfect Coquet," motivated by a desire "to render him more assiduous, and at the same time to regain that Liberty for her own Heart, which she found the Inclination she had to him above all other Men, was beginning to enthral."[100] Unfortunately, though, "[h]ow successful soever this way of Proceeding may sometimes have been found; it was far from answering the End *Amasina* proposed by it; and instead of rendering *Palamon* more submissive than he had been, made her appear to him every way less worthy of Respect."[101] She plays hard to get with Palamon, "contriv[ing] to be always Abroad at those Hours when she expected him to visit her."[102] She also flirts with "Beau *Trifle*, a Creature whose Conversation was shunn'd by every Woman of Prudence, to romp with her before his Face; and in fine, did even a Violence to her own Inclinations as well as to her Reputation, only to make Trial how far the Love *Palamon* had for her would compel him to bear."[103] Ultimately, Palamon feels compelled to "take [his] everlasting Leave," and Amasina finds herself "look[ing] back with Wonder and Detestation on those Irregularities which had deprived her of him," doomed to dissatisfaction in love for the rest of her life: "poor *Amasina* suffers her Bloom to wither in secret Repinings and unavailing Repentance, her Affliction heavier to be borne by the Endeavours she makes to conceal it."[104]

Though, according to Haywood's portrayal, Amasina is certainly blamable for her own misfortunes, the sensitivity with which the Female Spectator presents her story nevertheless invites readers to feel compassion for her, to relate to her, and to learn important lessons from the consequences of her behaviors. While narrating Amasina's story, the Female Spectator refers to her as "Poor unthinking Lady!"[105] The effect of this exclamation is not to belittle or demean Amasina. Rather, it cues readers to feel for her. Haywood encourages readers to recognize the circumstances that have motivated Amasina's behavior—in particular, a society that does not tend to expect women to think and that regularly encourages women to engage in trivial occupations:

> After all, no young Lady, if she thinks at all, can think the indulging herself too much in the modish Diversions of the Age will ever be agreeable to any Man, whose good Opinion it is worth her while either to inspire or preserve: Or can she answer it to her Reason that she takes more Pains to engage the idle Flatteries of a few unmeaning Coxcombs, than the solid Praises of Persons of Virtue and good Sense![106]

"Too many I fear," the Female Spectator observes, "are in *Amasina*'s Case, and for the Gratification of a Whim of a Moment's Duration, have sacrificed what would have made the Happiness of their whole Lives."[107] Taken together, these passages imply that, deprived of education and experience, most women of the time are disinclined to "think at all" and that they are prone to profound suffering as a result.

A reformed coquette herself, the Female Spectator speaks from personal experience and, thus, knows well how a life full of "modish Diversions" tragically prevents deep, authentic, and long-lasting intimacy (with men "whose good Opinion it is worth her while either to inspire or preserve"). She invites her readers (particularly, her female readers) not to castigate or ostracize Amasina, as a modest witness who makes his observations at a distance might, but rather to put themselves in her place and, thus, to feel as well as to learn from her pain: "By this Example young Ladies ought to be warned how dangerous it is to sport with the Affection of a Man of Sense."[108] The phrase "poor Amasina" recurs throughout the Female Spectator's narration of her story, and with each appearance the reader is prompted to be sensitive to, and affected by, her suffering so that they might become personally attuned to the learning opportunity her story affords.

Haywood's Sensitive Philosophy

In this final section of the chapter, I aim to show how Haywood, in *The Female Spectator*, indicates that such sensitivity not only enables individuals to learn important life lessons and to develop important skills of self-regulation but also to become adept philosophers. Thus, Haywood shows that, as sensitive witnesses, the Female Spectator and her followers become capable of asserting their own philosophical authority and of transforming their femininity from an impediment and barrier into an advantage that grants them access to domains that are otherwise inaccessible to them. The Female Spectator shares her brother's devotion to philosophy: "But true Philosophy as well as Religion will shew us better Things:—It will not only teach us the Nature and

Excellency of our Being, but also teach us how to avoid all such Inclinations as have any Tendency towards degrading its native Dignity, by throwing a Resemblance, or any way levelling us with the inferior Creation."[109] As this passage shows, like her brother, the Female Spectator believes that to practice philosophy is to improve the quality of one's life. Combined with religion, the "true Philosophy" Haywood here implies is that which "deals with the principles of human behaviour; the study of morality; ethics[;] . . . practical or proverbial wisdom; virtuous living" and which entails "[r]ational inquiry or argument."[110] However, this is not the only form of philosophy that Haywood practices and seeks to foster in *The Female Spectator*. As is the case for her brother, empirical philosophy—"the pursuit of scientific knowledge by means of observation and experiment"—is a key component of the Female Spectator's method of research.[111] Not only does she employ "Spies" to "extend [her] Speculations"; she also relies heavily throughout the periodical on her own personal experiences and direct observations.[112]

The Female Spectator does not simply practice empirical philosophy herself. In a series of issues inspired by correspondence ostensibly submitted by a reader identified as "Philo-Naturæ," she explicitly discusses empirical philosophy and recommends it to her readers.[113] In these issues, Haywood focuses specifically on the empirical study of the natural world, presenting it as a highly worthwhile pursuit, particularly for women. In fact, she indicates that empirical philosophy can and should provide a vehicle by which women, drawing on their capacity for sensitive witnessing and their close association with the natural world, may assert themselves intellectually and publicly alongside men. As is the case with all of the correspondence Haywood incorporates into her periodical, it is unknown whether the letters addressed by Philo-Naturæ were in fact composed by an actual reader or whether Haywood authored the letters herself. What is clear is that she enthusiastically approves of their sentiments. She provides his initial submission with a glowing introduction, referring to its author as "ingenious," even stating that "our Society takes a particular Pleasure in publishing [it]; as it agrees so exactly with our own Sentiments, and is what we would wish to say ourselves upon the same Occasion."[114] She goes on to welcome further submissions from him.[115] The primary aim of Philo-Naturæ's correspondence is to encourage women to pursue the educational opportunities afforded by "the Study of *Natural Philosophy*."[116]

He argues that such empirical philosophy is uniquely valuable to women because it enables them to fulfill their female duties while they pursue pleasure and education. According to Philo-Naturæ's portrayal, women's lives are intrinsically amenable to natural philosophy, and it can therefore be easily practiced within the female domain. In fact, Philo-Naturæ posits a close and significant association between women and specimens that natural philosophy commonly investigates. Thus, through the mouthpiece of Philo-Naturæ, Haywood portrays women as intimately connected to the objects of their investigation and claims such connection to be a strength rather than a weakness. Of caterpillars he writes:

> Even those worms which appear most despicable in our Eyes, if examin'd into, will excite our Admiration:—To see how in those little Creatures Bodies are cased in Bodies:—How, when one Form grows withered and decayed, the happy Insect has another in Reserve, and, shaking off the old, appears again in all the Freshness and Vigour of Youth:—What would a certain Lady, often taken Notice of in your Essays, and many other antiquated Beauties, give, they had the same Power?[117]

Philo-Naturæ here forges an anthropomorphic connection between "worms" and women that implicitly pivots on the similar littleness of their "Bodies." Inviting Haywood's female readers to identify with caterpillars, he not only encourages them to take a special interest in the "little Creatures" but also demonstrates the commonplace notion that women and natural philosophy make a good pair.[118] As his description of "the Ladies who keep Silk-worms" attests, natural philosophy is only a small step away from female "Finery": "They will tell you those pretty Creatures, from whose Bowles so much Finery is derived, after having finished their Work, erect themselves little Tombs, such as I have mentioned, and then revive in Butterflies in order to propagate their Species."[119] Philo-Naturæ thus brings natural philosophy home to Haywood's female readers, making it seem familiar and easily incorporated into their daily lives. Moreover, as he establishes a close association between women and their entomological specimens, he encourages them to feel for and with their specimens much as Haywood encourages readers to feel for and with her human subjects, even prompting them to imagine themselves having "the same Power" of regeneration as the caterpillars he invites them to observe. Thus, like the empirical philosophy the Female Spectator practices throughout the rest of the periodical, the natural philosophy Philo-Naturæ promotes is distinguished by sensitivity rather than Baconian modesty. Their personal intimacy and material continuity with their specimens, which their female gender helps enable and reveal, give them special access to empirical detail and, thus, to knowledge. It is as if Philo-Naturæ advises Haywood's readers to be like caterpillars so that her readers may come to know them.

Through her inclusion of and engagement with Philo-Naturæ's correspondence, Haywood demonstrates the potential for empowerment that sensitive witnessing holds for women. Though Philo-Naturæ does not go so far as to imagine Haywood's female readers attending meetings of the Royal Society or becoming members, he nevertheless does invite them to participate in this exclusive assembly through print: "As Ladies frequently walk out in the Country in little Troops, if every one of them would take with her a Magnifying Glass, . . . [t]hey would doubtless perceive Animals which are not to be found in the most accurate Volumes of Natural Philosophy; and the *Royal Society* might be indebted to every fair *Columbus* for a new World of Beings to employ their Speculations."[120] That Philo-Naturæ refers to women as "fair *Columbuses*" only highlights the way in which this passage opens the door for women to step outside femininity, even to engage in an act of figurative transvestism.[121] Though he elsewhere expresses a desire to keep women in their place, confining them to the domestic domain by making that domain more pleasant, here he expresses a hope that women might venture into more masculine locales. Thus, he portrays women not simply as delicate "Enliveners of Society" for men but also as determined explorers whose philosophical curiosity, once ignited, could defy conventional gender boundaries and even become worthy of public acknowledgment. It is no wonder, therefore, that Philo-Naturæ fears that fellow men may resent his encouragement of female philosophers: "I shall find little Forgiveness from my own Sex to persuade those Enliveners of Society to any thing which would deprive us of their Company for any long Time."[122]

As the Female Spectator herself proceeds to follow Philo-Naturæ's advice during a trip to the country, she provides an example of how women may pursue this more visionary aspect of Philo-Naturæ's recommendations. Haywood and her fellow female spectators are met with bad weather when they venture to "a Country Seat, belonging to our President, the excellent *Mira* . . . furnished with Microscopes, and other proper Glasses, in order to make those Inspections [Philo-Naturæ] recommends," and are therefore "obliged . . . to stay, for the most part, within Doors."[123] Thus, from the outset of their trip, their opportunities for exploration seem destined to be radically limited. However, the female spectators make the most of their rural retreat and manage to conduct a number of quite detailed empirical investigations in spite of the limits the weather imposes upon them. Indeed, one soon forgets that their investigations have been constrained by "continual Rain" and "poisonous Dew."[124]

Like Philo-Naturæ, Haywood accommodates natural philosophy to the traditional female station. The most overt illustration of this accommodation arises when she portrays the snails the female spectators collect for empirical study as a "young Nursery."[125] Yet through acts of magnification she also demonstrates that empiricism has the capacity to redefine the female station. After having spent a few days observing their "Nursery" of snails, the female spectators are "convinced . . . that this Species of Insects, contemptible as it may seem, had in it sufficient to excite an Admiration of the alwise and beneficent Creator, who forgets not the smallest of his Works."[126] "But, methinks, I hear some People say," Haywood writes, "*Could they find no Objects more worthy their Attention, than Caterpillars and Snails?—Two Insects the most contemptible of any.*" Her response to this challenge is telling: "[N]othing made by God is in itself contemptible:—Wonderful are all his Works, and the *Behemoth* of the Land, or the *Leviathan* of the Sea, magnify not his Power and Wisdom more by their Strength; not the spotted *Leopard* of the Forest, or the fine Limbed *Antelope*, or the starry-plumed *Peacock* by their Comeliness and Beauty, than do these Insects, by the amazing Properties bestowed on each."[127]

According to Haywood, all of God's creatures "magnify" the "power" of God; "amazing properties" have been "bestowed" upon them equally. Thus, snails and behemoths are equally admirable, and Haywood's feminization of nature's "contemptible" trivialities suggests that the same is true of women and men. Women, like caterpillars and snails, only seem inferior: small, mundane, even loathsome. In reality, Haywood indicates, they are no less than any of God's other creatures, whether behemoth, leviathan, leopard, antelope, "starry-plumed Peacock"—or even man.[128] The female sphere may be infused with the trivial but loses its triviality as well as its inferiority when subjected to the magnifying lens of natural philosophy.[129] In the hands of the Female Spectator, natural philosophy equalizes the masculine and the feminine, the male and the female, and thereby helps to promote women's self-assertion. According to the transformative power of empiricism, fair philosophers should be seen not simply as helpmates but as peers of male virtuosi.

Though Haywood's portrayal of empirical philosophy is distinctively feminine and even feminist, certain key aspects of England's new philosophy help justify Haywood's push for gender egalitarianism in the context of Baconian empiricism. For instance, though he might not have supported Haywood's gendered application, Francis Bacon would have fervently agreed with her claim that it is "often, by the most trifling Accidents, [that] very great and important Discoveries have been made."[130] He shares her conviction that what is perceived as trivial should be approached with great seriousness. In fact, he contends that "the greatest obstacle" to scientific progress is that "familiar things of frequent occurrence do not arrest and hold men's contemplation."[131] As Bacon here attests, crucial to his method was a deep reflection on ("contemplation" of) those common and humble phenomena that typically escape one's notice. According to Bacon, mundane, mean, and diminutive objects offer particularly important opportunities for philosophical investigation. He explains: "As for mean or even foul objects for which (as Pliny says) we have to apologise, . . . just as the best perfumes are sometimes made from smelly things like musk and civet, so excellent light and information emerge from mean and dirty things."[132] He therefore asks his reader to "[a]ccept into this history, first, the commonest things which one might think it superfluous to put into writing because they are so familiar; then mean things, illiberal, disgusting . . . ; also, trivial, childish things . . . ; finally, things which seem to be excessively subtle, because they are of no use in themselves."[133] For Bacon, the common, "mean," and "trivial" offer special access to the secrets of nature—and acquire a certain "dignity" in doing so. He even asserts, alluding to Matthew 18:3, that "we must become again as children utterly"—immersing ourselves in the trifles of everyday existence—if we are to make progress in natural philosophy.[134] As the ode Abraham Cowley provided to serve as a preface to Thomas Sprat's *History of the Royal Society* (1667) attests, from Bacon philosophers learned "to press/ Into the privatest recess/ Of [Nature's] imperceptible Littleness."[135] Investigations of the mundane, mean, and diminutive in fact provided the bedrock of England's new philosophy.[136]

That Haywood perceives natural philosophy as a means by which women may assert themselves, pressing beyond what typically is expected of them, is reinforced by the fact that she herself employs it thus. The example the Female Spectator provides through her own philosophic practice confirms that Philo-Naturæ's "fair Columbus" fantasy is well-founded. The Female Spectator embodies this fantasy in an imagined interaction with one of the gentlemen who have taken it upon themselves to guide her group's rural exercises:

> The worthy Gentleman, I before mentioned, and who is a great Contemplator of the minute Works of Nature, told us, that this Species of the Caterpillars is of the *Camelion* kind, and changes its Hue according to the Weather:—If we had continued in the Country a little longer, I would certainly have made the Experiment, by keeping one of them in a Box, with some Earth, and the same Sort of Leaves on which I found it feeding; for though I am willing to pay a due Deference to the Judgment of that Gentleman, I am rather apt to believe the Colour of these Animals more owing to their Food than to the Air they breathe.[137]

Though this interaction is only proposed, it nevertheless demonstrates that the Female Spectator experiences a remarkably profound sense of self-confidence and authority in the context of—and I would argue as a result of—her philosophic practice. Here, as elsewhere in her narrative of her visit to the country, she expresses "Deference to the Judgment" of a male virtuoso. However, such deference is belied by her unwillingness to be convinced by this "worthy Gentleman's" "Judgment."[138] Drawing on her past empirical experiences while also proposing future experimentation, she feels perfectly capable of taking a discordant view of "this Species of the Caterpillars." The submissive tone that the Female Spectator adopts threatens to obscure her philosophic boldness, yet it is clear from what she says here that her deference only goes so far.[139] This passage demonstrates that the Female Spectator feels qualified not only to discourse with men but also to disagree with them on subjects of natural philosophy.[140]

In fact, I would argue that, according to Haywood, ladies are potentially even more capable of excelling in natural philosophy than gentlemen are because of their plentiful leisure time and their *feminine* modesty but, even more importantly, their sensitivity to the natural world and their consequent profound capacity for "Heavenly Compassion."[141] The Female Spectator presents "a certain Lady of Quality, now living in *Lancashire*" as an icon of such female philosophical superiority.[142] This lady "has spent many Years in the Study of Physic," and thus offers an example of the philosophical advantage women possess as a result of their having "many [hours] . . . to spare [for] . . . the Study of Natural Philosophy."[143] She also exemplifies feminine modesty as, according to the Female Spectator, it is because "heaven has blessed" her "Prescriptions," not because she has been motivated by her own brilliance or ambition, that her remedies have achieved "Success."[144] Here, as elsewhere in *The Female Spectator*, Haywood portrays the female philosopher as humble and deferential even as she is assertive of her own views. Initially, Haywood's introduction makes this unnamed lady appear as something of a marvel and thus anomaly: "where the Diseases have been judged incurable by the Faculty, the Patient has not only been relieved, but entirely freed from them in a short Space of Time."[145] However, the Female Spectator nevertheless concludes her description of the lady with an encouragement to her readers not only to "admire" her but also to seek to "emulate" her.[146] Such emulation would appear far-fetched if the lady truly was a marvel, but instead, according to the Female Spectator's description, her philosophical prowess is merely a manifestation of her femininity, albeit a particularly dramatic one. She excels as a philosopher physician not in spite of her femininity but, rather, precisely because of it:

> This excellent Lady would scarce be brought to forgive a Servant, who should delay one Moment to acquaint her when any afflicted Person stood in need of her Relief.—Nothing is more common than to see her quit her Table in the midst of Dinner, and when surrounded by her Friends, to run to some Cottage, and exercise this Heavenly Compassion to her *Fellow-Creatures*, though in the most abject Station, and languishing under the most loathsome Ailment; and often has she, in the Dead of the Night, forsook her Bed, and mounted her Horse, without waiting till the Coach could be prepared, wholly regardless of Hail, Rain, Thunder, and Lightening:—In fine, no Time to her appears unseasonable,—no Weather unpropitious, when called to do the Work of Charity.
>
> Amazing *Commiseration*! and yet more amazing Fortitude and Courage![147]

As this passage shows, this female philosopher shares the Female Spectator's capacity to feel both for and with others—to commiserate with them and to be a sensitive witness to their experience. So accustomed is she to being ravished and penetrated by what she observes that even a hail storm cannot stop her. Sensitive witnesses both, for the Lady of Lancashire as for the Female Spectator, perception and observation are fundamentally sublime because they defy the boundary of the self. What's a hail storm to them? Also like the Female Spectator, her acts of commiseration are informed and enabled by a strong sense of fellow feeling for her "Fellow-Creatures." That the Female Spectator depicts these acts as instances of "Heavenly Compassion" emphasizes their significance as well as their selflessness, even suggesting that they refuse divisions between self and other. Thus, like the Female Spectator's, this female philosopher's acts of commiseration evoke Hume's theory of sympathy, which is, as we observed above, founded on a profound sense of "resemblance," and even continuity, with others.

In fact, the Female Spectator's depiction of this female philosopher shows that Haywood's sense of continuity between self and other is not restricted to human creatures but, in the tradition of Behn, Cavendish, and Hutchinson, encompasses all of the material world. In a fascinating article on *Fantomina*, Helen Thompson suggests that Haywood evokes Epicurean materialism (while previewing the work of Judith Butler) as she eschews Locke's theory of possessive individualism and instead portrays the subject as constituted and thus enabled by performance. As Thompson argues, by the end of *Fantomina*, the heroine appears as simply an "indifferent" pronoun, a neutral Epicurean body.[148] I would argue that much the same is true of this female philosopher from Lancashire. The Female Spectator introduces her to the reader after she and her female philosopher companions have, due to a violent thunderstorm, felt compelled not only to halt their astronomical investigations but also to retreat to a less exposed area of the building:

> the Corruscations of the Elements, instead of abating, became more outrageous, and several dreadful Claps of Thunder, accompanied with Lightnings that seemed to dart from every Quarter of the Heavens, filled us Women with such Terrors, that it was not in the Power of the Gentlemen to inspire us with Courage enough to continue in a Place where we imagined ourselves more exposed to Danger than in one where we should see or hear less of it.[149]

Immediately after describing the terror she and her female companions experienced, though, the Female Spectator characterizes their decision to retreat to safety as informed by a false presumption that the building will protect them: "So greatly do the Senses sometimes prevail over the Judgment."[150] Implicit here is that the Female Spectator and her female companions have allowed the terror that their sensory experience of the storm has provoked to overwhelm their ability to make reasonable and informed choices. In reality, as the Female Spectator proceeds to observe, even the strongest building is no match for a lightning strike:

> For, alas, if those Agents of Destruction were commissioned to strike us, where could we be safe?—Though hid in the Rocks, or in some Cavern in the Bowels of the Earth, there should we be found. . . . Common Observation . . . informs us, that Lightning is of that subtle penetrating Nature, that it can pierce through the thickest and most solid Bodies; we cannot therefore, when we reflect, hope any Protection from Walls composed either of Brick or Stone, yet in our Fright we run to them for Shelter, and are apt to accuse those of Presumption who, in Truth, are only more masters of Reason than ourselves.[151]

Here, Haywood joins the other female author-philosophers that this book explores as she portrays the "solidity of Bodies" as an illusion and emphasizes the porosity, malleability, and transcorporeality of matter.

It is within this context that the Female Spectator introduces the lady philosopher of Lancashire who, she explains, offers an "Example . . . that true Piety and a strong Faith can enable us to throw off all Delicacies and Fears, and venture, in a good Cause, all the warring Elements have Power to inflict."[152] So "amazing" is the lady's "commiseration" with those who are suffering—so sensitive is she to their plight—that even a threatening thunderstorm does not prevent her from braving the elements and, thus, subjecting herself to its "penetrating Nature."[153] This lady is far from in denial about her vulnerability to, and thus entanglement with, material forces. On the contrary, I would argue that, according to Haywood's portrayal, it is precisely her profound awareness of such entanglement that provokes her to risk her own safety. In this scene, this lady philosopher feels not only with those who "languish . . . under the most loathsome Ailment[s]" but with all of the material world. Pelted with hail and rain, jolted by thunder and lightning, she surrenders herself to her own material embeddedness. Thus envisioning a profoundly capacious "Heavenly compassion," this scene offers yet another instance in which the Female Spectator advocates feminine embodiment and sensitivity as philosophical ideals preferable to masculine modesty. Indeed, I would argue that this lady philosopher of Lancashire serves as a philosophical icon for Haywood. If only all philosophers (female as well as male) could be so pious, so full of faith, so brave.

Haywood's Skepticism

As this chapter has shown, whereas Addison and Steele promote Baconian modest witnessing in *The Spectator*, Haywood promotes sensitive witnessing to readers of *The Female Spectator*. In so doing, she allows her female readers to transform their perceived close identification with matter from a weakness into a strength. However, it is also important to acknowledge that, although Haywood portrays material entanglement as favorable to philosophical endeavors, she also strongly cautions against becoming overly caught up in such entanglement. To use the imagery of the Lady Philosopher of Lancashire anecdote as an illustration, Haywood encourages her readers to allow themselves to be "pelted with hail and rain" in the service of empiricism, yet she also shows them how to avoid being swept away by such storms—how to steel themselves from becoming subsumed by material stimulation. She wants them to experience material sublimity but not to be overcome (or undone) by it. The Lady of Lancashire surrenders but does not succumb to the storm. In fact, the imagery of her "mount[ing] her Horse, without waiting till the Coach could be prepared" evokes power, strength, determination—"amazing Fortitude and Courage!" Much like the Female Spectator herself, she is "engrossed" in and entangled by matter but is so steadfast in her will, purpose, and sense of self that she maintains her internal composure even as the tempestuous environment discomposes her.[154] I would argue that, for Haywood, skepticism must accompany sensitive witnessing if such composure is to be maintained.

The protagonist of Haywood's *Fantomina* is not as well prepared as the Lady of Lancashire. In fact, with *Fantomina*, Haywood presents an edifying foil for the Lady of Lancashire that cautions against the dangers of pursuing sensitivity without internal steadfastness. I would argue that, taken together, *The Female Spectator* and *Fantomina* advocate a particular variety of sublime knowledge-making that sets Haywood apart from the authors we have explored in previous chapters. Haywood's is a distinctively skeptical sublime characterized by a suspension of judgment and humility or even ambivalence about the possibility of establishing facts. Thus, surprisingly, Haywood joins her nemeses Alexander Pope and Jonathan Swift who, as James Noggle asserts, "associate sublime elevation with the baleful action of skepticism" as both "skeptical doubt and sublimity . . . chart the same intellectual motion, in which the mind tries to ascend to extraordinary levels of thought or perception only to fall into a doubtfulness that wipes out higher truths and common sense alike."[155] Truth claims are, for Haywood, provisional at best. She maintains that philosophers should be sensitive but must also simultaneously resist the temptation that sensitivity presents to become attached, fixated, and thus dogmatic—or what was, in the Enlightenment era, commonly referred to as "enthusiastic."[156]

Given the episodic eros of *Fantomina*, it can be easy to forget that the unnamed female protagonist's prime motivation is, in fact, neither sex nor love but curiosity.[157] As the first paragraph of the novella indicates, she initially decides to deceive Beauplaisir by donning disguises to seduce him because of her "wonder, that Men . . . should have Tastes so very Depraved" that they consort with whores (women "who come [to the theater] for no other Purpose, than to create Acquaintance with as many [men] as seem desirous of it").[158] Haywood proceeds to explain, "This excited a *Curiosity* in her to know in what Manner these Creatures were address'd."[159] Informed by this telling depiction of the protagonist's motivations, I would argue that, though her motivations ultimately become more amatory than empirical, we should view her, at the outset at least, as an empirical philosopher—specifically, a lady philosopher keen to know the world.

With *Fantomina*, Haywood presents a narrative arc that depicts the protagonist's downfall, and though superficially it appears as a downfall of a seductive, albeit naïve, young woman, I would argue that it is also—perhaps even more importantly so—a downfall of a philosopher. While this philosopher-temptress begins her escapades inspired by "curiosity," by the ninth paragraph of the novella, her prime motivation has shifted to love: "She loved *Beauplaisir*; it was only he whose Solicitations could give her Pleasure."[160] I would argue that, as she depicts this change in motivation, Haywood also depicts and warns against a danger that all philosophers must take care to avoid—namely, the danger of becoming fixated on, and even overcome (or in the case of *Fantomina* undone) by, their objects of study. Whereas the Lady of Lancashire remains constant in her philosophical purpose, the protagonist of *Fantomina* quickly abandons her philosophical purpose as her steadfastness becomes reduced to no more than a desire for Beauplaisir. In *Fantomina*, Haywood presents a parody of the kind of philosophical determination that the Lady of Lancashire represents: "with her Sex's Modesty, she had not also thrown off another Virtue equally valuable, tho' generally unfortunate, *Constancy*."[161] The Lady of Lancashire and *Fantomina*'s protagonist are both "constant." However, whereas the former is constant in her commitment to her philosophical purpose and the public service it inspires, the latter is constant in her lust for a rake, which ultimately results in her downfall. On the one hand, it would seem that, with *Fantomina*, Haywood seeks to warn women against dangerous desires for unworthy men. Though I would not challenge this view, I would argue that Haywood is particularly concerned in the work to educate female philosophers who, like *Fantomina*'s protagonist, find themselves motivated to explore the world by their curiosity. The lesson the novella provides closely corresponds to that which impels *The Female Spectator*: to be a good empirical philosopher, it is necessary both to become immersed within the material world but also to maintain a healthy skepticism while doing so in order to resist becoming swept away by it. In presenting this lesson, Haywood evokes Behn's *The Emperor of the Moon*, which also warns against being thus swept away. However, whereas Behn's lesson in her play is primarily directed at modest men, Haywood primarily directs hers at sensitive women. The protagonist of *Fantomina* would have had a much different fate had she managed to be more like the lady philosopher of Lancashire. In fact, had she done so, she could easily have served as one of the *The Female Spectator*'s spies, offering insight into what titillates men and what might inspire them to be constant lovers in spite of their obsession with novelty. *Fantomina*'s protagonist fails as a philosopher because she neglects to combine skepticism with her sensitivity.

Haywood presents the Female Spectator as a model of philosophical virtue as well, and as she does so, she shows the skepticism that is required to maintain such virtue and thereby avoid the fate of *Fantomina*'s protagonist. The Female Spectator's own story is one of reform. As we observed above, she begins the periodical with a recollection of her history as a "coquet."[162] Much like the protagonist of *Fantomina* during her pursuit of Beauplaisir, as a young woman the Female Spectator is driven by desire—"engross'd by a Hurry of promiscuous Diversions"—and constant to "Pleasure" rather than principle.[163] However, the ultimate fate of *Fantomina*'s protagonist and the Female Spectator are different. The former is shuttled off "to a Monastery in *France*" after having become pregnant by her lover.[164] The latter, though, has the opportunity to mature: "the too great Vivacity of my Nature became temper'd with Reflection" such that she becomes able "to *see* into the secret Springs which gave rise to the Actions I had either heard, or been Witness of—to *judge* of the various Passions of the human Mind, and *distinguish* those imperceptible Degrees by which they become Masters of the Heart, and attain the Dominion over Reason."[165] What allows her to develop such "sight," "judgment," and discernment? Haywood gives the impression that the accumulation of wide-ranging experience naturally gives rise to these capabilities: "A thousand odd

Adventures, which at the Time they happen'd made slight Impression on me, and seem'd to dwell no longer on my Mind than the Wonder they occasion'd, now rise fresh to my Remembrance, with this Advantage, that the Mystery I then, for want of Attention, imagin'd they contain'd, is entirely vanish'd, and I find it easy to account for the Cause by the Consequence."[166] The accumulation of experience causes "Wonder" and then "Mystery" to dissipate, and she evolves from credulous to incredulous, from gullible to suspicious. In other words, she becomes a skeptical, albeit still sensitive, philosopher.

I would argue that such skepticism infuses much of *The Female Spectator*. Throughout the periodical, Haywood refuses to accept truths just because they have been presented as such by other philosophers (even when the other philosophers are men and have authority). Moreover, she regularly criticizes credulity. We have already seen how the Female Spectator challenges her gentleman guide's analysis of the color-changing capabilities of caterpillars. This is indicative of her philosophical skepticism. Another and even more dramatic illustration of such skepticism arises during her expression of support for the theory of the "Plurality of Worlds." This theory was, as Haywood indicates, a source of controversy at the time of *The Female Spectator*'s publication. She first introduces it while narrating the telescopic investigations that the female spectators perform during their rural retreat. There is a lack of agreement on the topic among their gentlemen guides: "[T]hree of them were strongly for that System, and the fourth, who was of a contrary Opinion, had a very difficult Task to find Arguments which seemed of any Weight."[167] The Female Spectator proceeds to portray the rationales provided on both sides of the debate. As she does so, she demonstrates an obvious preference for the arguments offered as evidence for the theory ("System"), insisting that the gentleman who speaks on its behalf does so "with a good deal of Reason."[168]

Haywood returns to the controversy in a subsequent issue, prompted by a letter from Philo-Naturæ in which he speaks in favor of the theory. The Female Spectator responds to Philo-Naturæ as follows: "[O]ur Society are, above all, charmed with the convincing Defence he has made of an Opinion some Zealots and Enthusiasts so much cry down, and endeavour to explode as unchristian and fabulous: I mean that of Plurality of Worlds, which I never yet could hear any one good Reason to condemn."[169] Both here and, albeit less overtly, during her account of the gentlemen's debate on the topic, the Female Spectator shows a healthy skepticism as well as a willingness to make up her own mind on a much-disputed philosophic topic. Rather than remaining neutral on the subject, as a proper "deference" to male virtuosi would require, she takes a definitive stance in support of the Plurality of Worlds. Informed by her "Reason" as well as "*the Testimony of [her] own Eyes*," she firmly disagrees with the gentleman philosopher who rejects the theory.[170] In fact, labeling such philosophers "Zealots and Enthusiasts," she positions herself in direct opposition to them not only in terms of their position on the Plurality of Worlds controversy but also in terms of their philosophical methodology.[171] According to the Female Spectator, their philosophy is distinguished by irrational dogmatism, whereas hers is distinguished by reason and a healthy skepticism. The takeaway for her readers is: don't believe everything you hear, even if what you hear is presented by a gentleman virtuoso.

As Anton M. Matytsin and others have documented, Pyrrhonic skepticism was a crucial component of Enlightenment philosophy:

> By casting doubt over all received knowledge and questioning accepted methods for gathering true and certain information about the surrounding world, Pyrrhonian claims forced philosophers to formulate the principles and assumptions that they considered to be certain or, at the very least, highly probable. In attempting to answer the deep challenge of philosophical skepticism, these thinkers explicitly articulated the rules of attaining true and certain knowledge and defined the boundaries beyond which human understanding could not venture.[172]

In *The Female Spectator*, Haywood takes inspiration from such Pyrrhonian skepticism as she begins from a position of suspended judgment and seeks probability over certainty. However, experienced and expressed as it is within the context of her practice of sensitive witnessing, Haywood's skepticism assumes a distinctively radical quality. I would argue that, for Haywood, the modest philosophy advocated by Bacon and modeled by the Spectator is not skeptical enough. It is founded on a faith in the ability of man to purify himself of matter. As a female philosopher, Haywood believes herself to know better: humans are fundamentally material beings and should capitalize on, rather than credulously deny, this indisputable reality.[173] To emulate the Lady of Lancashire while engaging in sublime knowledge-making, as Haywood encourages her reader to do, is simultaneously to recognize and affectively experience our transcorporeal entanglement with matter and to appreciate that our knowledge of matter must always be embodied, situational, provisional and, therefore, humble. According to Haywood, whether you are a man or a woman, there is no such thing as a view from nowhere. In fact, the most reliable and thus best views are those from *somewhere.*

• • •

As we observed in chapter 1, Francis Bacon believed that, in order for philosophy to be accurate and reliable, practitioners had to cultivate modesty in order to be able to triumph over "the idols of the mind"—those human tendencies that have the effect of skewing philosophers' perception and analysis, thereby sabotaging the reliability and effectiveness of their practice. Recognizing the power of desire to spark zealotry and enthusiasm, Eliza Haywood was similarly concerned with the tendency of philosophers' observations and interpretations to be biased as a result of their natural inclinations. However, the strategy that Haywood advocates to prevent such bias is strikingly different from that recommended by Bacon. Whereas Bacon encourages philosophers to cultivate a "free and cleansed" mind and to transcend the material world, in *The Female Spectator* Haywood shows that she does not believe such an approach to be advisable or even feasible.[174] For Haywood as for the other author-philosophers this book addresses, interpenetrative transcorporeality is undeniable and, thus, the masculine modesty that Bacon and his followers promote is simply untenable. Moreover, according to Haywood, those who strive to cultivate Baconian modesty miss valuable knowledge-making opportunities. While the Spectator adopts a Baconian technique of detachment from matter, Haywood's Female Spectator instead immerses herself in matter while acknowledging the philosopher's own materiality. Moreover, Haywood presents sensitivity, rather than material detachment, as the best means for preserving philosophical accuracy and, thus, authority. For Haywood, so persistent is human desire that a philosopher would be naïve to think that he may avoid being swayed by it simply by striving to purify himself of, and thereby separating himself from, matter. That philosophers are themselves material beings makes such a belief all the more unwarranted. Haywood insists that preserving the integrity of philosophy is an ongoing and arduous process whereby one must remain devoted to philosophy as a principle even as one experiences the vicissitudes of material existence. To do so, one must remain skeptical and also consistently in tune to the mysteries and wonders of nature's touch.[175]

Like Margaret Cavendish, Haywood challenges the polarization of matter and mind, sense and reason. For Haywood as for Cavendish they are inseparable, and this inseparability demands a bipartite method of observation that dialectically combines sensitive perception and sensitive reason. Haywood goes even further than Cavendish, though, as she allows desire into her philosophy. As both the curious protagonist of *Fantomina* and the reformed coquette the Female Spectator, show, by fomenting and even themselves experiencing desire, philosophers gain special access to secrets of nature and society. The answer is not, therefore, as Bacon and the Spectator would have philosophers think, to purge oneself completely of desire but, rather, to temper it with a distinctively skeptical reason. Like Cavendish, Hutchinson, and Behn, Haywood advocates a philosophical method of sublime knowledge-making that is informed by the practice of sensitive witnessing. However, acknowledging the challenges that sensitivity can pose to philosophy, particularly when it manifests as desire, Haywood introduces radical skepticism into her method, suggesting that it has the capacity to protect philosophers from the temptation to allow their sensitivity and desire to metamorphose into attachment, zealotry, enthusiasm, dogmatism, and even obsession. Thus, the knowledge-making that Haywood entertains is characterized by a skeptical sublimity that is as circumspect about modern philosophy as it is passionate.

FOUR

A SCIENCE THAT CARES: CHARLOTTE SMITH

His heart
Was feelingly alive to all that breath'd.
—CHARLOTTE SMITH, *Beachy Head* (1807)[1]

AS THE SECOND HALF of the eighteenth century progressed, the practice of sensitive witnessing became encumbered by new associations and controversies. Mid-1700s Britain was distinguished by a "sentimental turn" that gave rise to a "cult of sensibility." As Alex Wetmore explains, this turn was "propelled by a conflux of ideological, economic, medical, religious and literary developments" and was characterized by "a widespread break with neoclassical rationalism."[2] Sensibility is defined by heightened emotional sensitivity. Consequently, to be a proponent of sensitive witnessing as a philosophical practice, one was forced to reckon with its relationship to sensibility as the term and the affective responsiveness it signified rose in popularity. Indeed, sensibility could be categorized as a form of sensitive witnessing and, certainly, the works this book has discussed thus far set an important foundation for the mid-century sentimental turn. However, the present chapter shows how late-Enlightenment British author-philosopher Charlotte Smith strove to preserve the philosophical authority that sensitive witnessing affords in spite of the increasing popularization of sensibility and sentimentalism.

Though there were many enthusiastic members of the cult of sensibility, it was nevertheless regularly derided as a pernicious fad that fomented extreme and even destructive behavior. Such criticisms only became more common and more biting as time went on. According to its critics, sensibility did not simply imply emotional sensitivity but, rather, emotional excess. As G. J. Barker-Benfield explains, [b]y the 1760s, . . . male writers were treating sentimentalization with a great deal of ambiguity, some of it clearly satire."[3] Sensibility had long been associated with sensitivity—"[t]he quality of being readily and strongly affected by emotional or artistic influences and experiences."[4] However, at mid-century, the term *sentimental* came to designate those who are "[c]haracterized by" or who "exhibit . . . refined and elevated feeling," and though initially this term was used "in [a] favourable sense," it would soon develop a negative connotation, signifying one who is "[a]ddicted to indulgence in superficial emotion."[5]

It is in this environment that author-philosopher Charlotte Smith rose to fame. As both a poet and novelist, Smith is regularly identified with sensibility and sentimentalism.[6] Reaching a particularly heightened pitch in her *Elegiac Sonnets* (1784) and her posthumously published blank-verse epic poem *Beachy Head* (1807), melancholy and intense feeling are hallmarks of both her poetry and her prose. However, Smith strove to preserve the philosophical authority that sensitive witnessing affords in spite of the increasing popularization of sensibility and sentimentalism. Like Haywood and Cavendish, as Smith develops and pursues methods of sensitive witnessing, she combines heightened affective experiences with reason and judgment. She also incorporates a new component into the practice: action. For Smith, like the other authors included in this book, sensitive witnessing entails an affective epistemology as well as an ontology of kinship. However, she maintains that it also implies an ethic—specifically, an ethic of care.[7] Consequently, she uses sensitive witnessing to inform and inspire her pursuit of social justice. As she does so, she emphasizes the radical aesthetic, feminist, and ecological dimensions of the practice while aligning it with the Sapphic sublime.

Smith's practice of sensitive witnessing is, like Haywood's, based on feeling for and with others and, therefore, is informed by an appreciation of one's continuity with, rather than distance from, objects of observation. Moreover, as is the case with the other authors this book explores, she invests in a distinctive method of sublime knowledge-making whereby the philosopher is perceived as entangled within, and inseparable rather than distinct from, their natural and social environments. As Kevis Goodman observes, in *Beachy Head*, Smith "intertwin[es] local subjects and global system—a system in which ecological, economic, and political elements intertwine."[8] I argue that such "intertwining" inspires and informs the ethic of care that Smith promotes for the sake of the public good. Finally, Smith maintains that, for sensitivity to enable sound philosophy, it must be paired with reflective reason. Writing not only after the cult of sensibility had taken hold but also at the time of the French Revolution, she was well aware of the dangers of rampant feeling. However, whereas some predictably promoted masculine modesty and the emotional detachment it entails as a means by which to counteract the excesses of sensibility and revolution, like the other female philosophers in this book, Smith was critical of masculine modesty. Not only did she believe it to be based on a delusion of humanity's separation from nature. Attuned to its tendency to promote dehumanizing reification, she also viewed it as an instigator of, and means of rationalizing, human cruelty and suffering. In this chapter, I argue that in the culminating work of her career, *Beachy Head*, Charlotte Smith developed and promoted a philosophical practice of sensitive witnessing as an alternative to the conventional practice of empirical modesty. To ensure its integrity as a philosophical and ethical practice, Smith insisted not only that sensitive witnessing must produce action but also that it should be imbued with humility and skepticism, pursued via a mixed method (combining poetry and empiricism), and reinforced by corroboration of evidence and reason. In this way, she succeeded in distinguishing her sensitive empirical practice from the self-indulgent, superficial, and philosophically unreliable expressions of sensibility that had, by the end of the Enlightenment era, become popular.

Charlotte Smith's Literary Empiricism

By the time she set to work on *Beachy Head*, Smith's investment in empirical philosophy was already well established. In fact, writing and empirical philosophy were intimately intertwined across Smith's career. She was herself an amateur botanist, and the study of plants regularly found its way into her writing.[9] A letter that she wrote to her publishers in 1797 demonstrates how closely related botany and writing were for Smith. In it, she outlines her desire to produce, in collaboration with her sister, a collection of "botanical drawings, . . . *one* to illustrate *each* of Linneas's orders."[10] She proceeds to explain that "when [she has] done [writing] twenty or thirty pages, [she] sometimes go[es] to [botanical] drawing by way of relaxation when other's would walk, or converse, or play at cards." Having "given up all society whatever," presumably either motivated by a desire to focus on her writing or simply by a preference for solitude, Smith seeks variety and "relaxation" from plants.[11] Viewed as important components of female education throughout much of the eighteenth and nineteenth centuries, both botany and drawing were common pastimes for English ladies.[12] However, the fluidity with which Smith oscillates between botany and literary composition is striking, offering a biographical context for corresponding oscillations in her literary works between botanical observation and poetic description. As this letter along with her verse suggest, botany and poetry are a natural pair for Smith.

In her educational writings as well as in *Beachy Head*, Smith provides her readers with opportunities to attain scientific knowledge while encouraging them to develop their own empirical practice. It is therefore in these works that the nuances of the empirical methods Smith employs and promotes may be discerned. Take, for instance, the scene from *Rural Walks* (1795) in which Mrs. Woodfield teaches her daughter Henrietta and her niece Caroline about orchis flowers:

> MRS. WOODFIELD: See how Henrietta is exploring the hedge-rows for flowers. Oh! she has found some treasure already!
>
> HENRIETTA: Mamma, I have found such an odd and pretty flower, and so sweet—sweet as an orange blossom—Do tell me the name of it?
>
> MRS. WOODFIELD: This is the white sweet smelling orchis; and is of the same genus of plants as those purple ones you gathered some days ago, and the paler lilac-coloured one, marked so beautifully with brown and black lines, which you found in the meadow this morning. If you look upon the down among the short turf, and on the chalky soil, you will find the orchis, whose flower resembles a bee, and another that represents a fly. Oh! Henrietta is fled after them already; she will certainly be a botanist![13]

This passage is typical of Smith's educational works in how it seeks to spark her readers' empirical curiosity while also providing them with lessons in natural history.[14] Here as throughout the rest of the work, Mrs. Woodfield and Henrietta serve as models to be emulated by Smith's readers, and they learn as Henrietta does. In this passage, Smith teaches her readers about the orchis genus and its variability. In other similar passages, they learn about snow-drops, black hellebores, mezerian, Daphné cneorum, pilewort, ranunculus, dormice, squirrels, nightingales, auricula, tulips, lily of the valley, lilacs, double blossom peaches, double blossom cherries, scorpion siennas, barberry, laburnum, etc.[15] As she states in the Preface to *Rural Walks*, with her educational works, Smith seeks to provide her readers with moral education:

> I have confined myself to what are called *les petites morales*. To repress discontent; to inculcate the necessity of submitting cheerfully to such situations as fortune may throw them into; to check that flippancy of remark, so frequently disgusting in girls of twelve or thirteen; and to correct the errors that young people often fall into in conversation, as well as to give them a taste for the pure pleasures of retirement, and the sublime beauties of Nature.[16]

Nevertheless, learning in natural history and empirical philosophy remains a constant priority as well. In fact, the empirical study of the natural world regularly provides the occasion for "les petites morales." For instance, it is in a botanical discussion of roses that she encourages her readers to seek "natural pleasures" by avoiding "artificial wants" and finding beauty in simplicity.[17] It is after having just discussed what makes a flower "wild" that she teaches them to care for the poor.[18] Pursuing a freely associative style that mimics her characters' peripatetic explorations of rural settings, Smith weaves opportunities to study and appreciate natural history throughout her educational works, demonstrating the value she accorded such learning.

Smith's poem *Beachy Head* also offers instruction in natural history and empirical philosophy, though its genre and style suggest a more mature and well-educated intended audience than her more overtly educational writings. Whereas the latter are addressed primarily to children (and secondarily, to their mother-teachers), *Beachy Head* appeals to a learned adult audience consisting of both women and men. Another significant difference between *Beachy Head* and her educational works is that whereas the latter were fairly conventional and expressive of broad contemporary trends, the former was pioneering and significant as a literary accomplishment. As Smith herself acknowledges in the preface to *Rural Walks*, "So numerous and so excellent are the books which have been written for the use of Children and Young Persons, within a very few years, that, on the great duties of life, nothing can, perhaps, be added, which is either new, or which can be addressed to them in any new form."[19] It is not surprising that Smith sought to capitalize on this trend in publishing, given the pressure she felt to earn enough money through her writing to be able to support her large family.[20] In fact, she appears to have been at least somewhat financially motivated in her composition of *Beachy Head* as well. Though we know little about Smith's motivations for writing the poem, as it was published posthumously and without an author's preface, one oblique mention of it appears in Smith's surviving correspondence: "I had just set down to a work which will bring me 2 or 3 hundred pounds."[21] As the editor of her correspondence, Judith Phillip Stanton, observes, "[p]robably here CS refers to *Beachy Head*, which eventually brought her £235 but was published posthumously."[22] Though this comment suggests that the promise of financial benefit was one of Smith's motivations for composing *Beachy Head*, its form, style, and substance indicate that a desire for literary accomplishment was also an important impetus for its composition. There is now a general consensus among scholars that *Beachy Head* offers a testament to Smith's literary genius and shows the author at the top of her literary game. A topographical epic in blank verse, replete with taxonomic footnotes, and suggestive of various prominent literary precursors, *Beachy Head* could hardly be more ambitious for a poet, and particularly for a female one.[23]

In addition to being her greatest work, *Beachy Head* offers valuable insight into Smith's engagement with and relationship to empirical philosophy. From the poem's first paragraph, Smith's interest and expertise in natural history feature prominently. Her inclusion of an abundance of footnotes—many of them quite extensive—providing Linnaean taxonomy as well as other empirical details and references explicitly designates the poem as philosophical. The first Linnaean note appears at line 23 as the third note in the poem, "The terns, and gulls, and tarrocks, seek their food": "Terns. *Sterna hirundo*, or Sea Swallow. Gulls. *Larus canus*. Tarrocks. *Larus tridactylus*."[24] Similar such taxonomic notes appear throughout the poem. They are evocative of, and likely inspired by, Erasmus Darwin who provides similar taxonomic notes in his *Loves of the Plants* (1789) and *Temple of Nature* (1803). Smith also regularly cites geological observations and theories in her notes. See for instance the second note of the poem: "Alluding to an idea that this Island was once joined to the continent of Europe, and torn from it by some convulsion of Nature." This note corresponds to the passage in the poem that describes:

> the strange and awful hour
> Of vast concussion; when the Omnipotent
> Stretch'd forth his arm, and rent the solid hills,
> Bidding the impetuous main flood rush between

The rifted shores, and from the continent
Eternally divided this green isle.[25]

Though as Anne D. Wallace observes, "neither the poem's explicit statements nor its figurative implications consistently align it with any specific geological theory," Smith nevertheless situates *Beachy Head* amidst contemporary geological debates, which the founding of the Geological Society of London in 1807 helped to fuel.[26] Through such notes, Smith demonstrates a deep familiarity with contemporary controversies in natural philosophy and an expertise in some of the most important works of natural philosophy current at the time.

The poem's notes do not simply indicate Smith's broad familiarity with philosophical works. They also attest to her own philosophical authority, which she expresses not only by referring to contemporary philosophical debates but also by challenging various accepted philosophical practices and theories. As illustrated by her regular citation of natural philosophers such as Linnaeus, Gilbert White (author of the 1789 *Natural History and Antiquities of Selborne*), and several others, her knowledge is informed by extensive reading. However, it is also informed by careful empirical study. Take, for instance, her footnote about Wheatears: "These birds are extremely fearful, and on the slightest appearance of a cloud, run for shelter to the first rut, or heap of stone, that they see."[27] Or, see her footnote about the Yellow Wagtail: "*Motacilla flava.* It frequents the banks of rivulets in winter, making its nest in meadows and corn-fields. But after the breeding season is over, it haunts downs and sheepwalks, and is seen constantly among the flocks, probably for the sake of the insects it picks up."[28] Such notes demonstrate an intimate acquaintance with the ecological diversity of Beachy Head, which has been achieved by careful and sustained empirical practice. She draws on this practice as she engages with and sometimes criticizes other philosophers' observations. For instance, in a note identifying the Fly Orchis (*Ophrys muscifera*), Smith comments, "Linnæus, misled by the variations to which some of his tribe are really subject, has perhaps too rashly esteemed all those which resemble insects, as forming only one species, which he terms *Ophrys insectifera*. See *English Botany*."[29] Smith here takes issue with Linnaeus's overgeneralization of orchids that mimic insects by placing them under the *Ophrys insectifera* species category, supporting her counterargument with a citation of James Sowerby's *English Botany*, which separates orchids that mimic different insects into different species as

Smith does: Bee Orchis, Fly Orchis, Spider Orchis, Butterfly Orchis.[30] In another footnote, she challenges Gilbert White: "Square holes cut in the turf, into which a wire noose is fixed, to catch Wheatears. Mr. White [*History of Selbourne*] says, that these birds (*Motacilla aenanthe*) are never taken beyond the river Adur, and Beding Hill; but this is certainly a mistake."[31] Here again, Smith provides what would appear to be a clarifying detail—"these birds . . . are never taken beyond the river Adur, and Beding Hill"—but then immediately counteracts such clarification by characterizing the information as "a mistake."

Thus, though Smith expresses admiration for Linnaeus in *Beachy Head*, attesting to the usefulness of his binomial system, she nevertheless complicates the elucidating effect of his taxonomy. Throughout *Beachy Head*, she consistently offers multiple names for the same specimen, thereby countering Linnaeus's goal of achieving identificatory consistency and implicitly promoting a more heteroglot method of identifying specimens. Thus, in spite of her admiration for him, she indicates that his taxonomical method is limiting, or at least insufficient in and of itself. See for instance her footnote for "gray choughs": "*Corvus Graculus*, Cornish Choughs, or, as these birds are called by the Sussex people, Saddle-backed Crows, build in great numbers on this coast."[32] In this footnote, as one name simply leads to another name, the poem and the birds it describes are imbued with heteroglossia. According to Smith's portrayal in this footnote, Linnaean taxonomy is not definitive but, rather, simply one among various ways of identifying specimens. Smith's "or" in this footnote indicates a certain agnosticism about what the right way is to identify "gray choughs": one possibility leads to another—from "gray choughs" to "*Corvus Graculus*" to "Cornish Choughs" to "Saddle-backed Crows." Similar such passages juxtaposing Linnaean binomial nomenclature with multiple other forms of identification recur throughout the poem.

Another similar instance appears in the footnote to her image of a "night-jar, chasing fern-flies":

> I remember only one instance in which the more remarkable, though by no means uncommon noise, of the Fern Owl, or Goatsucker, is mentioned. It is called the Night Hawk, the Jar Bird, the Churn Owl, and the Fern Owl, from its feeding on the *Scaraboeus solstitialis*, or Fern Chafer. It is called Goatsucker (*Caprimulgus*), from a strange prejudice taken against it by the Italians, who assert that it sucks their goats; and the peasants of England still believe that a disease in the backs of their cattle,

> occasioned by a fly, which deposits its egg under the skin, and raises a boil, sometimes fatal to calves, is the work of this bird, which they call a Puckeridge. Nothing can convince them that their beasts are not injured by this bird, which they therefore hold in abhorrence.[33]

Here again, Smith provides multiple names (ten total) for a specimen—in this case, the European Nightjar—separating them with *or*s and *and*s. Linnaeus's nomenclature is relegated to a parenthesis. Though the reader learns more about the people who name the nightjar than the nightjar itself through the multiple names Smith provides, their accumulation nevertheless has a significant effect on the reader's impression of taxonomy. For one, it helps contextualize and, thus, historicize the Latin name Linnaeus assigns to it: the genus *Caprimulgus* combines *capra*, which means "goat," and *mulgere*, which means "to milk." Consequently, it would seem that Linnaeus's choice of name is informed by a fictional legend, which Smith here attributes to the Italians. It also complicates, if not undermines, the authority of the name Linnaeus provides for the bird. Though his nomenclature may promise to help regularize the identification of the European Nightjar, Smith nevertheless presents it as merely one name among many. There is nothing inherently superior about the Linnaean name. In fact, the only real difference between it and the Italian name—"goatsucker"—is that it is in Latin. Thus, Smith suggests the arbitrariness of Linnaean taxonomy. Yet again, she pursues a heteroglot approach to classification that expresses her own philosophical authority while de-throning Linnaeus even as she conveys an appreciation for his taxonomic approach.

"Very vain is Science' proudest boast"

Though Smith expresses admiration for prominent empirical philosophers throughout *Beachy Head*, to recognize that she repeatedly complicates or challenges these philosophers is to recognize that she approaches their observations and discoveries with an intensely critical eye. Her admiration for them in no way entails blind adoration. In fact, I would argue that, much like the other authors this book has discussed, Smith fundamentally challenges their empirical methods, specifically taking issue with their reliance on masculine modesty. In fact, I would contend that *Beachy Head* is a polemic of empirical philosophy that proposes a better way of observing and discovering, a better way of doing natural philosophy and natural history, than had become common in mainstream philosophical circles across the Enlightenment era. While generally she does not pose this fundamental challenge directly or explicitly, the local challenges she presents to philosophers' observations and discoveries have a cumulative effect that suggests, in a subtle way, a broad dissatisfaction with their empirical methods. Such subtlety allows her to maintain her feminine deference even as she counters her male philosophical counterparts. However, that Smith is intent on criticizing current empirical practices becomes overt in a comment she makes in the middle of *Beachy Head*:

> Ah! very vain is Science' proudest boast,
> And but a little light its flame yet lends
> To its most ardent votaries; since from whence
> These fossil forms are seen, is but conjecture,
> Food for vague theories, or vain dispute.[34]

This statement is surprising in a poem that is so clearly influenced by, and indebted to, contemporary philosophers. In this one statement, though, Smith undercuts their authority and, in so doing, further complicates her reliance on them. Her main criticism here is of their vanity and pride. In their ardency, Smith suggests, they wrongly put their faith in "conjecture," "vague theories," and "vain dispute."[35] In actuality, according to Smith, "science" "lends" "but a little light" even to its most devoted practitioners. Philosophical modesty has, Smith suggests, gone to philosophers' heads. So lofty have they become in their perceived purity, detachment, aloofness, and transcendence that they have become prone to overestimating the value of their discoveries.[36]

To counteract such arrogance, brought on by philosophical modesty, Smith encourages philosophical humility as a preferable alternative, expressing and modeling it in various ways throughout *Beachy Head*. Take for instance her apostrophe to Ambition:

> Hither, Ambition come!
> Come and behold the nothingness of all
> For which you carry thro' the oppressed Earth,
> War, and its train of horrors.[37]

Though this passage does not relate directly to philosophy, I would suggest that it nevertheless has philosophical implications, particularly as it appears less than thirty lines after the passage about "Science' boast" quoted above. Together, both passages encourage not only philosophical skepticism but also humility. If "all" is "nothingness," then what can philosophy hope for?—to what can it aspire? No wonder it can "lend" only "a little light." Smith suggests that, in science as in war, "ambition" must be curbed. In fact, she goes so far in modeling and promoting skepticism and humility in *Beachy Head* that she challenges not only other philosophers' observations but also, at points, her own. For instance, in a footnote about a "night-jar, chasing fern-flies," Smith admits: "It was this bird that was intended to be described in the Forty-second Sonnet. I was mistaken in supposing it as visible in November; it is a migrant, and leaves this country in August."[38] Though Smith testifies to the philosophical value of empirical classification in *Beachy Head*, she nevertheless also consistently recognizes the limits of such value. All empirical philosophers, herself included, are prone to imprecision and error. Thus, skepticism and humility are, for Smith, crucial to philosophical authority.

I do not mean to suggest that Smith was unique in advocating philosophical skepticism and humility; however, the emphasis she places on these dimensions such that they infuse her entire poem is significant. As we observed in chapter 3, skepticism was an important component of the new philosophy. Even Francis Bacon manifested skepticism and humility. Although he was profoundly optimistic and ambitious in his vision for the "Great Instauration," he recognized that it would take a long time for its "harvest" to ripen enough to be ready to reap: "I do not run off like a child after golden apples, but stake all on the victory of art over nature in the race; nor do I make haste to mow down the moss or the corn in blade, but wait for the harvest in its due season."[39] The ultimate goal of the new philosophy was, for Bacon as for his followers, no less than a complete control of nature, and such control was predicated on a complete understanding of nature. However, as confident as Bacon was in his Great Instauration, he insisted that its culmination resided in a distant future; consequently, he embedded philosophical humility in his empirical method. The new philosophy was, from the beginning, viewed as a progressive pursuit:

> I have no entire or universal theory to propound. For it does not seem that the time is come for such an attempt. Neither can I hope to live to complete the sixth part of the Instauration . . . , but hold it enough if in the intermediate business I bear myself soberly and profitably, sowing in the meantime for future ages the seeds of a purer truth, and performing my part towards the commencement of the great undertaking.[40]

Bacon here attests that it is prudent for a philosopher to be satisfied with having made strides in the right direction. Progress towards "purer truth" is enough. In fact, it is all that can be hoped for in the current "time." Thus, Bacon himself advised philosophers to be humble in their hopes for accomplishment.

Humility and skepticism persisted as virtues alongside ambition throughout Enlightenment-era British philosophy.[41] As the following passage from Richard Pulteney in 1781 demonstrates, informed admirers of Carl Linnaeus, even as such admirers expressed a profound admiration for his system:

> No system yet invented can stand a rigorous examination through all its parts, and Linnæus was, perhaps, better acquainted than any other man with the defects of his own. The study of nature on scientific principles, notwithstanding the manifold improvements of later years, may yet justly be considered as in its infancy, and all arrangements hitherto proposed, have, in their turns, given way to others. At present the system of Linnæus possesses the advantage of a general superiority in the public approbation: how long it may enjoy this pre-eminence, time only can discover; in the mean while, it would be a more agreeable employment, to endeavour to strengthen its basis, supply its deficiencies, and candidly correct its errors, than to *object* to those anomalies and imperfections, which will most likely be ever inseparable from *artificial* arrangements.[42]

Though Pulteney's recognition of Linnaeus's "pre-eminence" generally informs his publication, he nevertheless feels compelled in the work's preface to acknowledge the philosopher's "deficiencies," "errors," "anomalies," and "imperfections" and to invite his reader to "correct" him where possible and necessary. Smith shares Bacon's and Pulteney's collaborative and progressive understanding of empirical philosophy, and she embodies humility and skepticism as she herself seeks to correct and thereby improve on contemporary philosophers' empirical observations while counteracting the "boastfulness" to which she believed they had become prone.

Smith also takes inspiration from conventional philosophical practices as she invests in the importance of the corroboration of evidence. In his discussion of how "assent," and thus reliable "judgement" of "matter[s] of fact" may be legitimately achieved, John Locke emphasizes the necessity of such verification: "[W]here any particular thing, consonant to the constant observation of ourselves and others, in the like case, *comes attested by the concurrent reports of all that mention it*, we receive it as easily, and build as firmly upon it, as if it were certain knowledge; and we reason and act thereupon with as little doubt, as if it were perfect demonstration."[43] It is when, and only when, "propositions about particular facts" are "agreeable to our constant experience" and/or when they are "generally spoke of, (when mentioned by others,) as things found constantly to be so, and therefore not so much as controverted by anybody," that "we are put past doubt." Such "*probabilities* rise" so "near to *certainty*, that they govern our thoughts as absolutely, and influence all our actions as fully, as the most evident demonstration: and in what concerns us, we make little or no difference between them and certain knowledge: our belief thus grounded, rises to *assurance*."[44] When one person's observation is confirmed by those of others, then its truthfulness becomes affirmed. It is precisely this collective approach to truth and knowledge that empirical philosophers of the Enlightenment pursued. From its inception, the new philosophy was portrayed as a collaborative enterprise. "The materials for the understanding," according to Bacon, "are so widely scattered that we need to have agents and merchants (so to speak) seeking and collecting them from every corner."[45] Moreover, it was commonly believed that, if the truth claims produced by the new philosophy were to be well founded, they had to be informed by a wide range of empirical testimony. Consequently, the Royal Society felt compelled to invite contributions from a broad spectrum of contributors. Sprat explains that it "embrac[es] all assistance": "All Places and Corners are now busie and warm about this Work; and we find many noble Rarities to be every Day given in not only by the Hands of learned and profess'd Philosophers: but from the Shops of *Mechanicks*; from the Voyages of *Merchants*; from the Ploughs of *Husbandmen*; from the Sports, the Fish-ponds, the Parks, the Gardens of *Gentlemen*."[46] To achieve empirical reliability, corroboration of evidence from a wide range of sources would be required. Smith's citation of philosophical authorities to contextualize her own empirical observations throughout *Beachy Head* is indicative of her participation in such a collective approach to truth and knowledge.

However, though Smith takes inspiration from such characteristics of contemporary empirical philosophy—its reliance on collective affirmation as well as its investment in skepticism and humility—she extends these characteristics much further than was conventional and combines them with an *im*modest sensitivity. As we observed in chapter 1, in spite of the lip service it gave to the importance of broad participation in the study of nature, the

Royal Society was in fact exclusive and exclusionary. Not only were women prevented from joining; as Steven Shapin has demonstrated, members tended to be of a certain social class: "the gentleman" was Enlightenment England's "paradigm of the type of individual one could trust to speak truth."[47] It was still the case when Smith composed *Beachy Head* that, to be modest, one was expected to be a gentleman.[48] Smith was not even able to achieve acceptance as a colorist of botanical illustrations, which she attempted in 1797 by offering her services to Dr. James Edward Smith, the founder of the Linnean Society.[49] However, throughout *Beachy Head*, as Smith asserts her own philosophical authority by offering her own observations and by challenging those of other philosophers as in the passages analyzed above, she refuses to capitulate to a narrow view of who is capable of speaking truth. She consistently indicates that, as a serious philosopher herself, she is authorized to offer her own empirical observations. Moreover, she suggests that what she observes has significant potential for being valuable to the collective study of nature—even when, perhaps especially when, her observations prompt her to accuse Linnaeus of being "misled."[50] According to Smith's portrayal, that she is a woman seems in no way to compromise her ability to produce and communicate important discoveries about the natural world. Furthermore, much as she broadens the collective and collaborative dimension of empirical philosophy, so too does she broaden the skepticism and humility it demands. When she claims that it is "but a little light" the "flame" of "Science . . . lends / To its most ardent votaries," she indicates that the understanding ("light") it provides is minimal ("little").[51] This is quite a striking statement in such a scientifically oriented poem. I would even go so far as to suggest that Smith's skepticism here significantly qualifies, if not completely undercuts, the very science on which *Beachy Head* relies. In fact, I would contend that, with this statement, Smith indicates that, if one's goal is to see things clearly, to understand and thus come to know nature, one must extend one's methods of knowledge-making beyond science as it is typically practiced.

Smith's Sensitive Empiricism and the Sublime

This is precisely what Smith achieves in *Beachy Head* as she develops an innovative method of sensitive witnessing. Smith suggests that, as empirical philosophers preserve their masculine modesty by remaining removed from their objects of study, they miss important opportunities for observation, understanding, and knowledge-making. It is because of their modesty that their science "lends but a little light." To achieve a more illuminating empirical practice, Smith invites philosophers to abandon their modesty and instead pursue a sensitive method that entails closeness and intimacy rather than distance and aloofness, that draws on emotion as well as observation, and that recognizes the philosopher's place in their own empirical practice. Throughout *Beachy Head*, Smith encourages the philosopher to touch and to be touched by nature—physically as well as emotionally. To do so is to trade science's "little" flame for a high-powered lantern. Such encouragement begins as early as the first sentence of the poem:

> On thy stupendous summit, rock sublime!
> That o'er the channel rear'd, half way at sea
> The mariner at early morning hails,
> I would recline.[52]

The poem opens with what would appear to be a conventional posture for a sublime prospect poem—the speaker atop an awe-inspiring cliff.[53] With its first line, "On thy stupendous summit, rock sublime!," *Beachy Head* evokes, and encourages readers to expect, a conventional sublimity. According to Edmund Burke, in a section of his *Enquiry* devoted to "Vastness," "Greatness of dimension, is a powerful cause of the sublime."[54] Similarly for Immanuel Kant, though he maintains that it is "in ourselves and the attitude of mind that introduces sublimity into the representation of nature," he nevertheless insists, "*that is sublime in comparison with which all else is small.*"[55] In the tradition of Burke and Kant, the first line of Smith's *Beachy Head* associates sublimity with vastness ("stupendous[ness]"), portraying such vastness to offer a highly affecting occasion for aesthetic transport. However, with the beginning of the fourth line, "I would recline," Smith eschews convention as she portrays the speaker in physical contact with the cliff. As she does so, she proposes an alternative orientation to nature that entails affective intimacy and sensitivity rather than masculine modesty. From the beginning of *Beachy Head*, Smith's speaker models and promotes a tender and reciprocal engagement with the material world, which enables her philosophical practice of sensitive witnessing.

In *Beachy Head*, Smith is as innovative in her portrayal of the sublime as she is in her empirical practice. By the time Smith composes *Beachy Head*, prospect poetry and poetry of the sublime had become closely associated with distance.[56] Throughout this book, we have recognized the importance of distance to the masculine modesty required by Baconian empiricism. By the end of the eighteenth century, distance had become important within literature of the sublime as well. In fact, I would suggest that sublimity, as conventionally defined, came to rely on a form of masculine modesty akin to that demanded by empirical philosophy. Prominent theorists such as Burke and Kant associate aesthetic pleasure with physical distance. In fact, according to both philosophers, distance is a prerequisite for the sublime. For Burke, who portrays sublimity as evocative of pleasurable terror, distance sets aesthetic experience apart from quotidian, lived reality: "When danger or pain press too nearly," observes Burke in his *Philosophical Enquiry into the Origin of Our Ideas of the Sublime and Beautiful* (1757), "they are incapable of giving any delight, and are simply terrible; but at certain *distances*, and with certain modifications, they may be, and they are delightful as we every day experience."[57] Without distance, terror can only be horrifying and, therefore, not a source of pleasure.[58] Kant places similar emphasis on physical distance in his *Critique of Judgment* (1790): "We must see ourselves safe in order to feel [the] soul-stirring delight" of the sublime.[59]

By portraying her speaker in the first sentence of the poem to "recline" on the "sublime rock," Smith both sets a foundation for her unconventional empirical practice and rebels against conventional sublimity. In the poem's opening lines, she takes a distinctive approach to the sublime much as she will proceed to take a distinctive approach to empirical philosophy. Such intimacy and sensitivity persist throughout the poem, and as they do, they reveal themselves to be vital to Smith's philosophical as well as aesthetic practice. Another passage that similarly illustrates her investment in intimacy and sensitivity appears in the middle of the poem within another apostrophe to the landscape:

> Ah! hills so early loved! in fancy still
> I breathe your pure keen air and still behold
> Those widely spreading views, mocking alike
> The Poet and the Painter's utmost art.
> And still, observing objects more minute,
> Wondering remark the strange and foreign forms
> Of sea-shells; with the pale calcareous soil
> Mingled, and seeming of resembling substance.[60]

Though it is the speaker's "fancy" here that conveys the imagery in these lines, Smith's repetition of "still" emphasizes that the imagery is nevertheless directly informed by her experience, which lends it an immediacy while stressing its autobiographical origin. Such immediacy and autobiographical character immediately infuse this passage with a sense of tenderness and affective responsiveness. Moreover, as Smith portrays the speaker to move closer to the cliff and the geological specimens that inhabit it ("observing objects more minute"), such sensitivity only becomes heightened. Given the intricacy of the seashells' shapes, to be capable of "remarking" their "strange and foreign forms," she must inspect them from a close-up vantage point. Smith zooms in on the landscape in this passage, and as she does so, she portrays the poem's speaker to be physically engaged with the landscape. Thus, here again, Smith opts for physical as well as emotional closeness rather than aesthetic distance.

As Smith cultivates such closeness, she attests to its value both to poetic expression and to empirical philosophy, thereby implicitly challenging the conventional belief that distance is necessary for sublimity or for reliable philosophical discovery to be possible. Both Smith's poetic and philosophical practice are sensitive rather than modest, engaged rather than aloof, in orientation to their objects of exploration and representation. Even as Smith's speaker "beholds widely spreading views," presumably observing them from a distanced vantage point, she simultaneously "breathes their pure keen air," thus taking them into her lungs so that they literally become a part of her. The word "keen" here suggests that, as the air of these views penetrate her body via her breath, she is able to feel them vividly.[61] Breathing in the "keen air," the speaker appears intermingled with the landscape, much as "the strange and foreign forms / Of sea-shells" appear here as "mingled" with the earth ("the pale calcareous soil"). In fact, I would argue that, due to such transcorporeal intermingling, it is as if seashell, soil, and speaker all "seem of resembling substance." Speaker and poet are therefore included within the material "composite" that "constitutes" Smith's "geological poetics."[62]

Moreover, as Smith portrays the speaker as a longtime "lover" of these hills, she emphasizes that her closeness with them is not only physical but also emotional. According to the logic the syntax of these lines suggests, it is in fact because of her love of the hills that the speaker is able to portray them so accurately. They may "mock alike / The Poet and the Painter's utmost art." However, in loving them, she has come to observe them regularly ("still," "still," "still") and, thus, to know them intimately so that she is better prepared to portray them than other poets, painters, and even philosophers are. These "widely spreading views" may be sublime and, thus, ineffable, but Smith suggests that they nevertheless become accessible to her speaker because of her affection for them as well as her close-up investigation of them. In fact, her intimacy with the hills enhances their sublimity, for as the speaker explores them closely, their "views spread" even more extensively ("widely").

In the vast sublime that Edmund Burke and Immanuel Kant promote, views typically only spread outward across extensive landscapes. Smith acknowledges the sublimity of such outward visual expansion here, but her main focus is on a different kind of expansion—namely, an expansion inward, *into* the hills, the "sea-shells," the "pale calcareous soil." As Judith Pascoe has shown, Smith's *Beachy Head* evokes microscopy in the close attention it gives to various specimens throughout.[63] In this passage, Smith's microscopic gaze uncovers a plurality of worlds within the landscape such that it becomes extensive internally as well as externally. In spite of his insistence elsewhere that small things are incapable of stimulating sublimity, even Burke feels compelled to acknowledge,

> [A]s the great extreme of dimension is sublime, so the last extreme of littleness is in some measure sublime likewise; when we attend to the infinite divisibility of matter, when we pursue animal life into these excessively small, and yet organized beings, that escape the nicest inquisition of the sense, when we push our discoveries yet downward, and consider those creatures so many degrees yet smaller, and the still diminishing scale of existence, in tracing which the imagination is lost as well as the sense, we become amazed and confounded at the wonders of minuteness; nor can we distinguish in its effect this extreme of littleness from the vast itself. For division must be infinite as well as addition.[64]

It is precisely such "divisibility of matter" that Smith illuminates, and uses to cultivate sublimity, as her speaker "observes objects more minute" within the chalk cliffs of Beachy Head. As a sensitive witness who is intimate with the landscape, Smith's speaker is inclined to "push her discoveries yet downward," and in doing so gains privileged access to a more obscure aspect of the cliffs' sublimity. The reader experiences this access vicariously and, thus, is given a special opportunity to "become amazed and confounded at the wonders of minuteness" at Beachy Head—wonders that inspire a sublime feeling that is, I would contend, more intense than that inspired by the landscape's "widely spreading views" because it is more surprising ("amazing" and "confounding").

In *Beachy Head* as well as in her *Elegiac Sonnets*, Smith develops a "tactile poetics" in which touch coordinates with other senses to induce a highly palpable form of aesthetic transport. The recurrence of tactile imagery throughout her writings is indicative of her investment in sensitive witnessing as both a poetic and philosophical practice. Through this practice, she portrays her self, via her poetic persona, as both highly responsive to and intimately entwined with its environment. Nowhere in her works is this self more clearly apparent than in *Beachy Head*. From the beginning of the poem, Smith's persona is highly sensitive to the landscape and its inhabitants. Its sensuousness immediately evokes such an orientation. Take for instance, these lines from the poem's first paragraph, which help establish the poem's setting as well as the speaker's intimate relationship with it:

> From thy projecting head-land I would mark
> Far in the east the shades of night disperse,
> Melting and thinned as from the dark blue wave
> Emerging, brilliant rays of arrowy light
> Dart from the horizon; when the glorious sun
> Just lifts above it his resplendent orb.[65]

Here, the speaker recounts her experience of observing sunrise from the prominence of Beachy Head. The imagery is profoundly evocative and detailed, illustrating that she is both stimulated and inspired by its dazzling features. Or, see this passage in which, while reminiscing about her youth, the speaker explains,

> An early worshipper at Nature's shrine,
> I loved her rudest scenes— . . .
> And hedge rows, bordering unfrequented lanes
> Bowered with wild roses, and the clasping woodbine
> Where purple tassels of the tangling vetch
> With bittersweet, and bryony inweave,
> And the dew fills the silver bindweed's cups—
> I love to trace the brooks whose humid banks
> Nourish the harebell, and the freckled pagil;
> And stroll among o'ershadowing woods of beech,

Lending in Summer, from the heats of noon
A whispering shade; while haply there reclines
Some pensive lover of uncultur'd flowrs,
Who, from the tumps with bright green mosses clad,
Plucks the wood sorrel, with its light thin leaves,
Heart-shaped, and triply folded; and its root
Creeping like beaded coral.[66]

In centering botanical specimens that are known for rapid vining and growth, Smith here portrays the encounter between speaker and her environment as both intimate and reciprocal, attributing a kind of dynamism, even agency, to nonhuman nature. It entangles her in its "inweave," insisting on closeness. The speaker invites and enjoys its touch, strolling among the shadows, while feeling the cool of the shade "whispering" on her skin. She returns nature's touch as she "trace[s] the brooks" with her feet. The intimacy only deepens as she happens upon a "pensive lover of uncultur'd flowrs" plucking the wood sorrel whose leaves are "heart-shaped, and triply folded." It is as if the wood sorrel presents love letters from nature to humanity that are received not only by the "pensive lover" of flowers but also by the speaker and, vicariously, the reader of the poem. Here, not only does Smith portray the speaker as sensitive; nature is sensitive too. I would argue that, by portraying such an intimate intermingling between human and nature, self and world, Smith replaces the distance from nature entailed by the masculine modesty required by more conventional eighteenth-century approaches to aesthetic pleasure and empirical philosophy with an expansive ontology of kinship with nature. She calls on humans to perceive their bodies and selves as intertwined with, and highly sensitive to, the world around them such that the fate of self becomes inseparable from that of the other.

Smith's Mixed Method: Science and Poetry

Smith's philosophical method is, fundamentally, what would now be called a "mixed method." It mixes science and poetry, as well as empirical observation and feeling. As Smith pursues such combinations, she suggests that her mixed method is superior to those whose methods are more restrictive and homogeneous—more modest. *Beachy Head* is a testament to there being many ways of knowing and the value of making use of as many as possible. In her footnote for "cuckoo-flowers," she comments, "poets have never been botanists" after having corrected Shakespeare's description of "Cuckoo buds as being yellow [in *Love's Labour's Lost* V.ii.894]": "He probably meant the numerous Ranunculi, or March marigolds (*Caltha palusris)* which so gild the meadows in Spring. The Cuckoo flower is the *Lychnis floscuculi*."[67] It is precisely as a botanist-poet that Smith styles herself in *Beachy Head*, though, so the real message of this footnote is: poets have never been botanists *until now*.[68] While Smith does not exactly claim superiority over Shakespeare, she does imply, through her own example, that there is a value to poetry informed by botanical knowledge: as a botanist, she is capable of achieving greater empirical accuracy and precision than other authors. Throughout the poem, she regularly demonstrates an expansive and highly developed knowledge, not only of botany but of natural history more generally (with particular expertise in birds and geology). She also implies, however, that there is a value to empirical philosophy informed by a poetic outlook as well. For one, according to Smith, such an outlook allows a philosopher to enhance their ability to describe what they observe, which is, according to Smith, particularly important given nature's sublimity. See for instance, these lines in which she observes the awe-inspiring view that the cliff affords:

> Beneath the shadow of a gnarled thorn,
> Bent by the sea blast from a seat of turf
> With fairy nosegays strewn, how wide the view!
> Till in the distant north it melts away,
> And mingles indiscriminate with clouds.
> . . . Yet hence the distant range
> Of Kentish hills, appear in purple haze;
> And nearer, undulate the wooded heights,
> And airy summits, that above the mole
> Rise in green beauty; and the beacon'd ridge
> Of Black-down shagg'd with heath, and swelling rude
> Like a dark island from the vale[69]

In her footnote to these lines, she observes, "Description falls so infinitely short of the reality, that only here and there, distinct features can be given."[70] Smith thus suggests that, due to its sublime obscurity, this view thwarts the "naked way of writing" that characterizes conventional empirical description.[71] However, even as she attributes a certain ineffability to the landscape, Smith's poetic language and imagery nevertheless manage to convey its magnificence as well as its empirical details. By mixing science and poetry, both of which are, after all, typically distinguished by careful attention to physical features, Smith enhances her ability to depict what she observes.

With such a mixed method, Smith again deviates from conventional empirical philosophy as she devises her own distinctive empirical practice. Though there was precedent for such a mixed method, most prominently in the work of Erasmus Darwin, it nevertheless remained an exception to use poetic description to conduct and convey serious empirical philosophy.[72] As Tita Chico observes, "from the early days of the Royal Society, . . . figurative associations were admonished, particularly with regard to reporting scientific discoveries."[73] Abraham Cowley's "Ode to the Royal Society," first published as a prefix to Thomas Sprat's *History of the Royal Society of London, for the Improving of Natural Knowledge* (1667), offers a particularly memorable depiction of the Enlightenment empiricist's determination to avoid "artful" language:

> His candid style like a clean stream does slide,
> And his bright fancy all the way
> Does like the sunshine in it play;
> It does like Thames, the best of rivers, glide,
> Where the god does not rudely overturn,
> But gently pour the crystal urn,
> And with judicious hand does the whole current guide.
> 'T has all the beauties nature can impart,
> And all the comely dress, without the paint, of art.[74]

From the beginning, the new science had been characterized by a plain style—what Cowley here refers to as "candid" and "clean." This style was expressive of and conducive to the masculine modesty to which Bacon and his followers aspired. In fact, the 1663 statutes establishing the Royal Society explicitly outline the importance of maintaining a method of empirical presentation that eschews literariness: "In all Reports of Experiments to be brought into the Society, the matter of fact shall be barely stated, without any prefaces, apologies, or rhetorical flourishes; and entered so in the Register-book, by order of the Society."[75] In *Beachy Head*, Smith rebels against this directive, suggesting that matters of fact are more fully expressed when science and art are combined than when science is purified of art.

Though she relies heavily on taxonomy and conventional scientific description, Smith also regularly recognizes their limitations. Indeed, throughout *Beachy Head*, on its own, empirical philosophy often obscures at least as much as, if not more than, it clarifies. This is particularly true for a reader who is not well-versed in Linnaean taxonomy and/or Latin, as would have been the case for many of Smith's readers. However, even for one fluent in Latin and/or familiar with Linnaean classification, such a taxonomic web is not necessarily clarifying. Whereas Erasmus Darwin's Linnaean footnotes are designed to be explanatory and, thus, to satisfy the reader's curiosity, thereby enhancing the meaning of the poem, Smith regularly challenges, or at least circumscribes, the empirical value of Linnaean taxonomy. As a point of contrast, take Darwin's description of a Canna flower:

> First the tall CANNA lifts his curled brow
> Erect to heaven, and plights his nuptial vow;
> The virtuous pair, in milder regions born,
> Dread the rude blast of Autumn's icy morn;
> Round the chill fair he folds his crimson vest,
> And clasps the timorous beauty to his breast.[76]

The footnote Darwin provides for this passage is as follows: "*Canna*. l. 39. Cane, or Indian Reed, One male and one female inhabit each flower. It is brought from between the tropics to our hothouses, and bears a beautiful crimson flower; the seeds are used as shot by the Indians, and are strung for prayer-beads in some catholic countries."[77] As is common throughout Darwin's *Loves of the Plants*, the footnote here serves an explanatory function in relation to the passage in the poem to which it corresponds. The poem's passage personifies the canna as a married husband and wife ("the virtuous couple), and the Linnaean footnote helps the reader understand why: each Canna flower is inhabited by "one male and one female." In other words, Cannas are hermaphrodites, with male and female reproductive organs occupying the same specimen. The footnote also helps the reader understand why the canna "dread the rude blast of Autumn's icy morn": it is "brought from between the tropics to our hothouses" and is therefore not able to withstand the English cold.

In contrast, Smith's Linnaean footnotes are rarely as satisfyingly explanatory. See for instance the footnote she provides for the lines "who there/ Gathers, the copse's pride, anémones,/ With rays like golden studs on ivory laid/ most delicate: but touch'd with purple clouds, / Fit crown for April's fair but changeful brow": "Anémones. *Anemóne nemorosa.* It appears to be settled on late and excellent authorities, that this word should not be accented on the second syllable, but on the penultima. I have however ventured the more known accentuation, as more generally used, and suiting better the nature of my verse."[78] Like Darwin, Smith here offers a figurative depiction of her botanical specimen, using both simile ("with rays like golden studs on ivory laid") and metaphor ("touch'd with purple clouds, / Fit crown for April's fair but changeful brow") to enrich and enliven her imagery. However, rather than providing any explanation to accompany the taxonomy as Darwin does, Smith instead chooses to let the taxonomical definition stand alone.[79] For a reader not fluent in Latin, its Linnaean term is void of meaning, and if it communicates at all, it does so by its sound, which offers a musical combination of consonance and assonance, rather than by its signifying capacity. Even for one fluent in Latin, though, the anemone's Linnaean classification is potentially more poetic than clarifying. *Nemorosa* signifies shady or wooded, whereas *anemóne* signifies wind or even daughter of the wind. In other words, the Linnaean classification is itself metaphorical and poetically suggestive. Smith emphasizes this quality by allowing it to stand alone without explanation. Thus, by providing the Linnaean taxonomy, Smith exacerbates rather than diminishes the obscurity of her image of anemones. That she proceeds in the footnote to acknowledge a lack of consensus regarding how the word *anemone* is pronounced only further enhances such obscurity. Even the flower's pronunciation is shrouded in mystery.

By thus augmenting the obscurity of the many specimens that *Beachy Head* explores, Smith further emphasizes the "littleness" of the "light" that empirical philosophy's "flame lends" and thereby provides further support for her skeptical and humble empirical practice. Even footnotes that seem definitive in *Beachy Head* assume a tentative status as they are juxtaposed with equivocal ones. See for instance her footnote for the line, "And the short turf is gay with tormentil," which is simply: "tormentil. *Tormentilla reptans.*"[80] Such succinct and ostensibly straightforward footnotes are common throughout *Beachy Head.* Taken alone, they would appear to be definitive and elucidating. However, because they appear alongside less definitive ones such as those discussed above, they too seem provisional at best. By challenging the authority of experts like Linnaeus and White, Smith encourages her readers to look with some skepticism at all of the empirical details her footnotes provide. Such skepticism prepares her readers to appreciate such details not simply as empirical truths but also, and perhaps more so, as contributions to the poem's literary effect. I would argue that in her simple taxonomic footnotes as in her more equivocal ones, the Linnaean taxonomy operates at least as much as poetry as it does empirical clarification, with consonance and assonance regularly contributing to its poetic effect. Throughout *Beachy Head*, Smith invites the reader to appreciate *the poetry* of Linnaean taxonomy.

It is true that, throughout the Enlightenment era, some fellows of the Royal Society would publish "artful" or poetic literary works that drew on their scientific knowledge, which seem to deviate from the modest and plain style promoted by the society. See for instance Henry Baker's sublime poem, *The Universe* (1734), in which Baker provides the following gloss for the passage, "What is this Earth, of which thou art so proud? / Lost and unknown, in the more glorious Crowd, / A Point it scarce appears": "Our *Glasses* discover innumerably more *Stars* than we can discern with the naked Eye; and still the better our *Glasses* are, the more we find out, lying beyond the other, and so on, for anything we know, indefinitely and inexhaustibly."[81] Many of the footnotes Baker includes in *The Universe* are, as this one is, directly informed by his empirical practice as a fellow of the Royal Society. However, as the work's title page makes clear, the goal of the poem is not to present empirical observations but, rather, "To restrain the Pride of Man." Baker's is an artful, poetic work informed by empirical philosophy but not itself an empirical work. In contrast, empirical observation and presentation are among Smith's main goals in *Beachy Head* and at points appear as her main priorities. This is a fine distinction but an important one. In *Beachy Head*'s opening lines, Smith's speaker "represents" and "marks," identifying depiction and observation as principal aims:

> I would recline; while Fancy should go forth,
> And *represent* the strange and awful hour
> Of vast concussion; . . .
> From thy projecting head-land I would *mark*
> Far in the east the shades of night disperse,
> Melting and thinned, as from the dark blue wave
> Emerging, brilliant rays of arrowy light
> Dart from the horizon; when the glorious sun
> Just lifts above it his resplendent orb.[82]

If we compare Smith's opening lines to Baker's, the contrast between their purposes becomes clear:

> Thy Works, Eternal Power by whom she sings!
> The *Muse* attempts, and tunes the sounding Strings:
> To Heav'n and Thee her Adoration raise,
> And form the Song devoted to thy Praise![83]

Smith's *Beachy Head* aims to observe ("mark") and depict ("represent") whereas Baker's *The Universe* aims to "praise."

Moreover, though Baker indulges in artful, poetic language in *The Universe*, in his empirical writings he is restrained. In these works, Baker aims to follow the plain style expected of fellows of the Royal Society, as the following passage from one of his papers in the *Philosophical Transactions* illustrates: "*Plate VI.* shews these curious fossil bodies at more than half their real bigness. They are only three in number, though there are four figures, one of them being drawn in two positions. They are evidently of a boney substance, made black, most likely, and rendered brittle, by some mineral steams or juices, though not corroded by them."[84] Though Baker takes inspiration from his empirical practice in his poetry, he nevertheless strives to keep his empirical practice itself free of poetic language.[85] In contrast, Smith uses her poetry to express and document her empirical practice. Smith's is a mixed empirical method whereas Baker's is a modest one.

Granted, as Chico demonstrates, even the most conventional empirical philosophers regularly could not help themselves from relying on figurative and thus "artful" language, particularly when conveying "something that the human eye cannot perceive."[86] For instance, in a depiction of the horse fly's eye that Royal Society fellow Henry Power includes in his empirical work, *Experimental Philosophy in Three Books* (1664), the author explains that "it looks like green silk Irish-stitch, drawn upon a black ground, and all latticed or chequered with dimples."[87] Here, Power resorts to simile to portray what he sees. Nevertheless, in spite of thus periodically resorting to figurative language, plain style remained the goal of properly modest Enlightenment empirical philosophers, and when they deviated from it, they resisted acknowledging having done so. As Chico explains,

> The concern was to prove that experimental accounts in no way embellished the details—that is, natural philosophers had to assure readers

> that accounts did not draw on the imagination in any overt or identifiable way. . . . [T]he practitioner of natural philosophy "endeavours rather to know, than to admire."[88] Intellectual mastery is captured by the term *know* and contrasted with the "Romantic" quality of admiration. The former comes across as a means to establish distance and difference between the observer and the observed. The latter conveys identification and a loss of self. Sprat's natural philosopher aims "to enlarge, and amend his *Knoweldge* . . . [and is] modest in his *Judgment*."[89]

Enlightenment era empirical philosophers were generally determined to maintain their philosophical as well as rhetorical modesty in order to preserve their authority. In *Beachy Head*, Smith presents an unconventional and unapologetic combination of empiricism and poetry while seeming perfectly satisfied "to know" and "to admire" at the same time. Though she asserts her "intellectual" authority, she declines to claim the "mastery" that her more conventional philosophical counterparts claimed. As we have already seen, humility is a key component of Smith's empirical practice. Moreover, whereas conventional empirical philosophy of the Enlightenment era "established distance and difference between the observer and the observed" as a means of substantiating the mastery, purity, and thus modesty of the philosopher, Smith's distinctive empirical practice entails intimacy and continuity between the observer and the observed as a means of substantiating her sensitivity and thus authority as a philosopher.

Smith's Science and the Sublime

As Smith mixes science and poetry, so too does she mix empirical discovery with the sublime and thus evokes the various other author-philosophers this book has explored who have also portrayed knowledge-making as a sublime experience. To know best, Smith's *Beachy Head* suggests, one should be willing to let nature in, to be emotionally and physically intimate with nature while also recognizing one's own kinship with materiality. Though there are various instances of sublime "soaring" in *Beachy Head*—for instance, "Thither aspiring Fancy fondly soars, / Wandering sublime thro' visionary vales"—Smith persistently brings the speaker's perspective back down to earth and water: "and on the opposing side / The early moon distinctly rising, throws / Her pearly brilliance on the trembling tide."[90] Such smooth progression from sky to sea or earth, in addition to Smith's recurring portrayal of the reflection of the sky by the sea and earth, refuses a separation between sublime soaring and earthly wandering.[91] In fact, throughout *Beachy Head*, sky, sea, and earth are all sublime and are also all united by their shared materiality.[92] I would therefore argue that Smith's is, from the beginning, a "material" as opposed to a transcendental sublime.[93] As such, it provides an apt occasion for her sensitive empirical method. See for instance this passage in which Smith speculates about the "strange and foreign forms / Of sea-shells" she finds "with the pale calcareous soil / Mingled":

> [S]urely the blue Ocean (from the heights
> Where the downs westward trend, but dimly seen)
> Here never roll'd its surge. Does Nature then
> Mimic, in wanton mood, fantastic shapes
> Of bivalves, and inwreathed volutes, that cling
> To the dark sea-rock of the wat'ry world?
> Or did this range of chalky mountains, once
> Form a vast bason, where the Ocean waves
> Swell'd fathomless? What time these fossil shells,
> Buoy'd on their native element, were thrown
> Among the imbedding calx: when the huge hill
> Its giant bulk heaved, and in strange ferment
> Grew up a guardian barrier, 'twixt the sea
> And the green level of the sylvan weald.[94]

In a footnote accompanying these lines, Smith remarks, "The theory here slightly hinted at, is taken from an idea started by Mr. White [Gilbert White (1720–93), *The History of Selbourne* (1789)]."[95] Once again, we here see Smith's embodiment of philosophical humility and skepticism in her empirical practice. Rather than offering a single theory, she offers several, employing a technique of surmise that refuses certainty.[96] Here too we see the mixed nature of her method. This passage is simultaneously scientific and poetic in its exploration of specimens and in its diction. Seashells are both empirical specimens ("bivalves," "volutes") and aesthetic artifacts ("fantastic shapes"). Furthermore, in spite of their smallness, they embody the sublime much as the vast rolling "blue Ocean" does. Through her imagery of the spiraling seashells, Smith invites us to experience awe at what their forms represent—namely, either the terrific geological shifts that their presence in the cliff potentially signifies or the tendency of "Nature" to "Mimic, in wanton mood" on land what exists in the sea. However, she also invites us to be transported by their "strange and foreign forms" themselves. They may be small, but they are in and of themselves awe inspiring—in their strangeness and mystery. In fact, I would argue that, in this passage, the spiraling seashell serves as an emblem of the material sublime, which unlike the transcendental sublime is readily inclusive of not only vastness but also smallness. Moreover, whereas the transcendental sublime is contingent on distance, the material sublime embodied by Smith's spiraling seashell invites a breaching of distance for to be able to observe and admire their "fantastic shapes," one must view them from a close-up vantage point.

I would argue that Smith's material sublime is best understood as a Sapphic sublime, for it not only entails intimate rather than distanced encounters between the observer and observed but also embodies a recognition of the materiality and consequent tenuousness of human existence—the fact that, as material beings, we are all engaged in a process of dissolution that, though tragic, unites us and thus allows us to commune with sublimity. It was not until the aesthetic philosophy of Burke and Kant that the sublime became fully masculinized, associated with patriarchal power, and distanced, abstracted.[97] In fact, in the earliest surviving theory of the sublime, first century CE *Peri Hupsous*, the author Longinus even cites a female poet, Sappho, who is distinguished by sensuousness and materiality as one of his examples of poetic sublimity:

> Now, as there are no Subjects which are not attended by some adherent Circumstances, an accurate and judicious Choice of the most suitable of these Circumstances and an ingenious and skillful Connection of them into one body, must necessarily produce the Sublime. For what by the judicious Choice, and what by the skilful [*sic*] Connection, they cannot but very much affect the Imagination.
>
> *Sappho* is an Instance of this, who having observ'd the Anxieties and Tortures inseparable to jealous Love, has collected and displayed them all with the most lively Exactness. But in what particular has she shewn her Excellence? In selecting those Circumstances which suit best with her Subject, and afterwards connecting them together with so much Art.[98]

Longinus then proceeds to quote a fragment by Sappho that describes the intense emotional condition she experiences as a result of her "jealous Love":

> Blest as th' immortal Gods is he,
> The Youth who fondly sits by thee,
> And hears, and sees thee all the while
> Softly speak, and sweetly smile.

'Twas this deprived my Soul of Rest,
And raised such tumults in my Breast;
For while I gaz'd, in Transports tost,
My Breath was gone, my Voice was lost.

My bosom glow'd; the subtle Flame
Ran quick thro' all my vital Frame;
O'er my dim Eyes a Darkness hung;
My Ears with hollow Murmurs rung.

In dewy Damps my Limbs were chill'd;
My Blood with gentle Horrors thrill'd;
My feeble Pulse forgot to play,
I fainted, sunk, and dy'd away.[99]

Longinus's use of Sappho's poem as an illustration of the sublime shows just how significantly Burke and Kant depart from his *Peri Hupsous*. The fact that she is a woman is, in and of itself, an important difference. Burke and Kant thoroughly deny women access to the sublime. As Ian Balfour explains,

> Longinus himself used the terms *kalos* (beautiful) and *hypsos* (sublime) as if they were interchangeable, much to the consternation of Burke. Yet as the eighteenth century wore on, the tendency was, increasingly, to separate as strictly as possible the sublime from the beautiful. In Burke one could learn not only that the sublime and the beautiful were categorically different from each other, but that each was aligned . . . with one sex rather than the other.[100]

In fact, Burke is unable even to allow a fictional female *character* into his sublime aesthetic. Burke cites Milton's Death as a quintessential example of sublimity, but he completely ignores Death's mother and close companion, Sin. Presumably, Sin's gender as well as her fleshiness preclude her from Burke's definition of the sublime.[101] According to Burke, in Milton's description of Death "all is dark, uncertain, confused, terrible, and sublime to the last degree":

The other shape,
If shape it might be called that shape had none
Distinguishable, in member, joint, or limb;

> Or substance might be called that shadow seemed,
> For each seemed either; black he stood as night;
> Fierce as ten furies; terrible as hell;
> And shook a deadly dart. What seemed his head
> The likeness of a kingly crown had on.[102]

Though Milton's Sin is certainly terrible, her form is not "uncertain" or obscure like Death's is. Death is abstract, vague, almost ethereal whereas Sin is exceedingly material, which in addition to her gender prevents her from being worthy of inclusion in Burke's portrayal of the sublime:

> Before the Gates there sat
> On either side a formidable shape;
> The one seem'd Woman to the waste, and fair,
> But ended foul in many a scaly fould
> Voluminous and vast, a Serpent arm'd
> With mortal sting: about her middle round
> A cry of Hell Hounds never ceasing bark'd
> With wide Cerberean mouths full loud, and rung
> A hideous Peal: yet, when they list, would creep,
> If aught disturb'd thir noyse, into her woomb,
> And kennel there, yet there still bark'd and howl'd
> Within unseen.[103]

Characterized by an investment in materiality and sensuousness ("Voluminous and vast"), Smith's *Beachy Head* evokes the material sublimity portrayed by Sappho and Milton's Sin, which is a variety of sublimity that Burke and Kant exclude in spite of the fact that Longinus celebrated it. As in Sappho's portrayal of her love for Phaon and in Milton's portrayal of Sin, the sublimity that Smith portrays in *Beachy Head* is dramatically embodied and thus experienced not primarily through imagination, as is the case in Burke, or reason, as is the case in Kant, but rather through palpable feeling whereby even the virtual or the distant feels proximate.[104] That Beachy Head the place has been reputed as a site for suicide since the seventh century lends further evidence to the Sapphic nature of the poem's sublimity, as legend has it that, inspired by her unrequited love for Phaon, Sappho committed suicide by jumping from a cliff into the ocean. Whereas Burke and Kant imply masculine modesty and transcendence as they maintain that distance and safety are necessary to the sublime, in contrast, the Sapphic sublime implies material entanglement not only as it suggests physicality, palpability, and transcorporeality but also as it is associated with real danger, physical harm, and even death.

Darby the Hermit

How such a material sublime informs and contributes to Smith's philosophical practice of sensitive witnessing is best observed in the conclusion of *Beachy Head*, which narrates the story of Darby, "a lone hermit" and an embodiment of sensitivity.[105] According to Smith's portrayal, and to quote this chapter's epigraph, Darby is "feelingly alive to all that breathes." [106] I would argue that, as such, he serves as a figure for the type of poet-philosopher Smith aspires to be. Consequently, though the poem is ostensibly unfinished (published in "an imperfect state" due to "the increasing debility of its author," according to its original editor), his story provides a fitting conclusion.[107] By the time Darby appears in the poem, he is already familiar, resembling as he does the sensitive speaker herself. In a footnote, Smith tells the reader that this character is based on an actual person: "a man of the name of Darby, who for many years had no other abode than this cave, and subsisted almost entirely on shell-fish."[108] While Darby is not a philosopher per se, he does practice meteorological empirical observation as he

> learn[s] . . . to augur from the clouds of heaven,
> And from the changing colours of the sea,
> And sullen murmurs of the hollow cliffs,
> Or the dark porpoises, that near the shore
> Gambol'd and sported on the level brine
> When tempests were approaching.[109]

There are other important similarities between him and Smith's speaker—her persona—in the poem as well. Not only are they both similarly solitary as well as inclined toward deep and capacious feeling. They are also both physically intimate with the landscape. As we have seen, Smith's speaker begins the poem "reclining" on the cliff at Beachy Head and, throughout, perceives the details of the landscape from a close-up vantage point, communing with its details as she experiences their sublimity and studies them empirically. Darby similarly communes with Beachy Head. He lives "[j]ust beneath the rock / Where Beachy overpeers the channel wave, / Within a cavern mined by wintry tides"—literally embedded within the landscape.[110] Thus, like Smith's speaker, Darby is emotionally as well as materially entangled with Beachy Head.

His death at the end of the poem only further emphasizes such entanglement:

One dark night
The equinoctial wind blew south by west,
Fierce on the shore;—the bellowing cliffs were shook
Even to their stony base, and fragments fell
Flashing and thundering on the angry flood.
At day-break, anxious for the lonely man,
His cave the mountain shepherds visited,
Tho' sand and banks of weeds had choak'd their way—
He was not in it; but his drowned cor'se
By the waves wafted, near his former home
Receiv'd the rites of burial.[111]

In this case, the imagery of irresistible entanglement between human and nonhuman nature Smith presents is associated with tragedy and death rather than the pleasure that infuses material entanglement throughout much of the rest of the poem. However, it nevertheless attests to humanity's intimacy with nonhuman nature much as the imagery of entanglement throughout the rest of the poem does. In fact, placed at the end of the poem, it portrays death as an apotheosis of humanity's entanglement with their environment. Consequently, though Darby's death is tragic, it is also affirming. For one, it lends support to the practice of sensitive witnessing that Smith promotes by portraying sensitivity—openness, receptivity, and vulnerability to one's environment—as a fundamental human condition, which is merely denied by modest philosophy. Moreover, it emphasizes that, creatures of nature ourselves, even the most solitary and "lonely" human is never truly alone for we are always mingled in and, thus, touched by and touching nature. I would assert that Darby's fate stands as a figure for all of our fates: in the end, transcorporeal beings that we are, we will return to, and become absorbed by, nature. Masculine modesty here appears as a distant and delusional fantasy. How can one assert one's distance from nonhuman nature when our intermingling with it is so profound and fundamental?

That the poem's last lines create the impression that the entire poem is, in fact, an epitaph for Darby reinforces his centrality to the poem as a whole. In its conclusion, the poem becomes an elegy, mourning the loss of a solitary sensitive witness gone too soon:

Those who read
Chisel'd within the rock, *these mournful lines,*
Memorials of his sufferings, did not grieve,
That dying in the cause of charity
His spirit, from its earthly bondage freed,
Had to some better region fled for ever.[112]

Whether intentionally or because the poem is unfinished, Smith declines to include a clear antecedent for "these mournful lines." Perhaps she had planned to insert a poetic epitaph for Hermit Darby at the end of the poem but was never able to do so due to her infirmities at the end of her life. However, as is, the poem presents *Beachy Head* itself as the antecedent of "these mournful lines" and, thus, as Hermit Darby's epitaph. We therefore ultimately discover that not only is the poem a paean to the practice of sensitive witnessing; it is also inspired by, and composed as a memorial for, sensitive witness par excellence, Darby. Consequently, Darby comes to serve as an icon and martyr for the practice of sensitive witnessing that the poem promotes. I would argue that, ultimately, the poem's conclusion conveys a two-part message that advises the reader: (1) to be sensitive like Darby is, and (2) that we are all like Darby, whether we are willing to admit it or not.

Sensitive Witnessing, Sensibility, and Reason

Though Smith never uses the term *sensibility* in *Beachy Head*, it nevertheless provides an important literary-historical context for the poem, and it is certainly relevant both to her memorialization of and identification with Darby and to her development of sensitive witnessing as a philosophical practice. To be "feelingly alive to all that breath'd" is to be endowed with sensibility. However, Smith's portrayal of Darby shows how she not only incorporates, and thus capitalizes on, but also significantly revises tenets of the cult of sensibility. In her engagement with sensibility, as in her engagement with empirical philosophy, Smith was an innovator. In more conventional expressions of sensibility, authors would regularly portray feeling as valuable in and of itself as "sensibility [became] a form of religion."[113] In contrast, throughout *Beachy Head*, Smith insists on the importance of integrating feeling with reason and action. Profound emotion is a key component of Smith's practice of sensitive witnessing, yet emotion alone is not enough: to be valuable it must be informed by reason, and beneficial action must be its end. To understand how

Smith integrates feeling and reason in *Beachy Head*, it is helpful to appreciate how she sets the foundation for such integration in her first major publication of poetry, *Elegiac Sonnets* (1784), which first established her as a poet of sensibility. According to Elizabeth Dolan, "Smith pioneered two techniques in the *Elegiac Sonnets* that became core characteristics of poetry in the emerging Romantic movement—the displacement of emotion on to nature, and the poetic tempering of emotion through rational thought."[114] The collection's association with sensibility is immediately apparent both in its title and in its preface, where Smith explains that the included poems have been inspired by "[s]ome very melancholy moments."[115] Reason's presence in the collection is not as immediately obvious; however, it is nonetheless significant, and her portrayal of it gives important clues to how she understands it.

Smith uses the term *reason* in various sonnets throughout the collection. Its first appearance is in the final quatrain of "Sonnet XXI. Supposed to be written by Werter." The sonnet concludes thus:

> Tow'rds the deep gulf that opens on my sight
> I hurry forward, Passion's helpless slave!
> And scorning *Reason's* mild and sober light,
> Pursue the path that leads me to the grave!
> So round the flame the giddy insect flies,
> And courts the fatal fire by which it dies![116]

This sonnet is first in a sequence of five written from the perspective of the melancholy and tragic protagonist, Werther, in Johann Wolfgang Goethe's 1774 epistolary novel, *The Sorrows of Young Werther*. Here, Smith juxtaposes reason and "passion," aligning the speaker with the latter. By the time Smith published her *Elegiac Sonnets*, not only had the cult of sensibility caught hold but so too had the cult of Werther. Extremely popular in England as in Germany, he was regularly portrayed, celebrated, and emulated as an icon of sensibility.[117] That Smith includes a sonnet sequence devoted to him in her *Elegiac Sonnets* is a marker of Smith's own status as such an icon. Like Werther, she regularly evinces a tendency to prioritize heightened feeling over reason and related concepts such as logic and practicality throughout *Elegiac Sonnets*. However, as this passage from the first poem in her Werther sequence shows, she expresses a strong ambivalence about such prioritization. Though she writes from Werther's perspective with a profound sense of understanding as well as kinship, she also bemoans his suicidal fate ("the path that leads him to the grave"). If Werther is a hero, he is a tragic one in Smith's portrayal (as he is in Goethe's). Admittedly, in these lines, there is a certain romantic allure to his status as "Passion's helpless slave." However, Smith nevertheless invites the reader to bemoan his refusal to follow "Reason's mild and sober light," a depiction of reason that closely corresponds to Hume's depiction of it as "a general calm determination of the passions."[118] In the final couplet, she unflatteringly analogizes Werther to a "giddy insect" who, behaving mechanically, "flies round the flame" until it succumbs to the "fatal fire."

Reason appears again in the third sonnet of the Werther sequence, "Sonnet XXIII. By the Same. To the North Star." In this sonnet, Smith analogizes reason to the North Star. I quote this poem in full here so as to appreciate the full significance of the analogy, which ensures that though reason is not named until the final couplet, it is nevertheless present throughout the entire poem:

> To thy bright beams I turn my swimming eyes,
> Fair, fav'rite planet! which in happier days
> Saw my young hopes, ah! faithless hopes!—arise,
> And on my passion shed propitious rays!
> Now nightly wandering 'mid the tempests drear
> That howl the woods and rocky steeps among,
> I love to see thy sudden light appear
> Thro' the swift clouds—driven by the wind along;
> Or in the turbid water, rude and dark,
> O'er whose wild stream the gust of Winter raves,
> Thy trembling light with pleasure still I mark,
> Gleam in faint radiance on the foaming waves!
> So o'er my soul short rays of *reason* fly,
> Then fade:—and leave me to despair, and die!

Here again it is Werther's inability (or refusal) to follow the guidance of reason that provokes his "despair" and his death. Smith employs chiaroscuro to evoke the contrast between reason, which Smith associates with "propitious rays" of light, and "passion," which she associates with "nightly wandering," "tempests drear," "swift clouds," "turbid water . . . o'er whose wild stream the gust of Winter raves," and "foaming waves." Though in his youth ("happier days"), reason, like the North Star, "saw his young hopes . . . arise," in his maturity, their "short rays" "fade," and as they do, he succumbs "to despair" and death. Though the Werther sonnet sequence affirms the power of passion, extreme feeling, and, thus, sensibility, it also attests to the importance and value of reason. Smith portrays reason as necessary to survival, particularly for those who, like Werther and herself, are prone to extreme feeling. She invites her readers to relate to and even admire Werther, but she also demands that we mourn his death. She suggests that, had he found a way to combine feeling with reason, he might not have felt compelled to commit suicide. Whereas Goethe resists such a combination—particularly, in the stark opposition he maintains between Werther and the character of Albert—Smith, in contrast, invites it.

This is not to say that, for Smith, reason is an easy quality to achieve or, for that matter, to endure. In "Sonnet XXXVIII," she personifies it as a tyrannical queen:

> Alas! these joys are mine in dreams alone,
> When cruel Reason abdicates her throne!
> Her harsh return condemns me to complain
> Thro' life unpitied, unrelieved, unknown!
> And as the dear delusions leave my brain,
> She bids the truth recur—with aggravated pain!

"Harsh" and "cruel," reason here appears as an instigator of suffering ("aggravated pain"). Upon the speaker's awakening from his dreams, she "bids the truth recur" and thus halts his "joys."[119] Similarly, in the final lines of "Sonnet LXX: On being cautioned against walking on an headland overlooking the sea, because it was frequented by a lunatic," Smith portrays reason as a curse:

> In moody sadness, on the giddy brink,
> I see him more with envy than with fear;
> *He* has no *nice felicities* that shrink
> From giant horrors; wildly wandering here,
> He seems (uncursed with reason) not to know
> The depth or the duration of his woe.

Here, Smith's speaker claims to "envy" the "lunatic" because, without reason, he is unaware of his own "woe" whereas, endowed with reason, her speaker is acutely aware of hers. Such portrayals of reason as severe, unkind, and burdensome contribute to the collection's reputation as a work of sensibility as they appear to suggest that Smith's allegiance is to feeling as opposed to reason.[120] However, I would argue that Smith's portrayal of both reason and feeling is more nuanced and complex. It is not that she eschews reason in *Elegiac Sonnets.* Rather, she recognizes the pain it can bring, especially for the sensitive witness. However, while she views reason as a source of pain, Smith nevertheless also insists on its value for, without it, the sensitive witness is in danger either of becoming a "lunatic" or, like Werther, of perishing altogether. Reason is necessary for the sensitive witness's very survival.

Sensitive Witnessing and the Public Good: An Ethic of Care

Smith emphasizes the importance of reason in *Beachy Head* as well. However, there, its value inheres more in its capacity to provoke virtuous action than in its ability to ensure one's survival. Its importance in *Beachy Head* becomes explicit in a passage in which Smith describes the harvesting of pearls by enslaved people in Asia:

> There the Earth hides within her glowing breast
> The beamy adamant and the round pearl
> Enchased in rugged covering; which the slave,
> With perilous and breathless toil, tears off
> From the rough sea-rock, deep beneath the waves.
> These are the toys of Nature; and her sport
> Of little estimate in Reason's eye;
> And they who reason with abhorrence see
> Man, for such gaudes and baubles, violate
> The sacred freedom of his fellow man—
> Erroneous estimate![121]

As this passage explains, it is "they who reason" who find the violation of "the sacred freedom of his fellow man" "abhorrent." It thus becomes clear here that reason has the power to enhance one's capacity to feel and, thus, to practice sensitive witnessing. Without reason, Smith implies, one may feel dismay at the suffering of the enslaved, but for one to be able to experience feeling that adequately registers (with appropriate "abhorrence") the pearl gatherer's anguish and the violation that has inflicted it, reason is necessary. Thus, this passage illustrates that, for Smith, sensitive feeling is best when combined with reason. According to Smith, sensibility and reason should go hand in hand.

I would argue that Smith's integration of feeling and reason closely corresponds to the theory of moral sense developed by Mary Wollstonecraft in her

1790 political pamphlet, *A Vindication of the Rights of Men*. Though scholars regularly characterize Wollstonecraft as one of the chief critics of sensibility, she nevertheless remained invested in the importance and value of feeling to virtuous action throughout her career. As Alex Wetmore explains, "the deeper one probes, the more deeply entangled Wollstonecraft's writing becomes with the literature and culture of sensibility, even as sentimentalism's limitations for women remain a constant and prevalent concern."[122] This passage from Wollstonecraft is particularly relevant to the relationship between feeling and reason that Smith promotes:

> Men who possess uncommon sensibility, whose quick emotions shew how closely the eye and heart are connected, soon forget the most forcible sensations. Not tarrying long enough in the brain to be subject to reflection, the next sensations, of course, obliterate them. Memory, however, treasures up these proofs of native goodness; and the being who is not spurred on to any virtuous act, still thinks itself of consequence, and boasts of its feelings. Why? Because the sight of distress, or an affecting narrative, made its blood flow with more velocity, and the heart, literally speaking, beat with sympathetic emotion. We ought to beware of confounding mechanical instinctive sensations with *emotions that reason deepens*, and justly terms the feelings of *humanity*. This word discriminates the active assertions of virtue from the vague declamation of sensibility.[123]

Here, Wollstonecraft emphasizes the importance of distinguishing "emotions that reason deepens" from "mechanical instinctive sensations," recognizing the importance of the former while portraying the latter as superficial and insignificant. She suggests that it is not enough to have "uncommon sensibility," for alone strong feelings ("sympathetic emotion") are fleeting and fail to result in anything but self-involved delusions of goodness that cause one to "boast of [their] feelings" without performing "any virtuous act." Wollstonecraft challenges those who are satisfied to be endowed *only* with "uncommon sensibility"—who believe feelings in and of themselves are enough. She thus implicitly criticizes the cult of sensibility here, which had developed the tendency to fetishize feeling without any consideration of how it might be made valuable to the public good. However, this passage demonstrates that it would be wrong to say that Wollstonecraft rejects sensibility altogether. Instead, she maintains that sensibility becomes enhanced and, thus, valuable when it is paired with reason. According to her portrayal, for emotions to be deepened by reason is for them to become subject to "reflection" and, thus, to be transformed into "the feelings of *humanity*" that motivate "active assertions of virtue." Reason thus rescues sensibility from solipsism as feelings themselves become perceived as collective ("of *humanity*") rather than, simply, as the sole property of the feeling self. Endowed with a heightened capacity for feeling and with reason, the individual develops a sense of obligation to act virtuously and, thus, in the service of the public good. I would argue that it is precisely such a productive combination of feeling and reason that distinguishes the practice of sensitive witnessing Smith develops across her career. For Smith as for Wollstonecraft, feeling alone is not enough. Virtuous action must be the ultimate goal, and for feeling to lead to such action, it must be deepened by reflective reason.

Smith similarly challenges the mechanical operation of feeling—what Wollstonecraft refers to as "mechanical instinctive sensations." As Smith observes in "Sonnet XXI: Supposed to be written by Werter," such mechanical operation makes one a "helpless slave" to "Passion."[124] It is precisely because Werther allows himself to become such a slave, Smith suggests, that he suffers a horrific fate. Like Wollstonecraft, Smith instead promotes a combination of passion, or "mechanical instinctive sensations," and reason so that reason may deepen emotions that warrant being deepened. Smith's speaker in *Beachy Head* embodies this combination, as her "abhorrence" at the pearl gatherer's enslavement shows, and so too does the poem's other example of the sensitive witness, Hermit Darby. Not only does he "acutely fe[el] / For human misery."[125] He also acts on these feelings, ultimately sacrificing his own life in the service of others. In Smith's portrayal (and also in local legend), Hermit Darby is a regular savior of imperiled mariners. As Smith's footnote explains, according to "the people of the country[,] . . . [h]e had often administered assistance to ship-wrecked mariners":

at night
He listen'd to the wind; and as it drove
The billows with o'erwhelming vehemence
He, starting from his rugged couch, went forth
And hazarding a life, too valueless,
He waded thro' the waves, with plank or pole
Towards where the mariner in conflict dread
Was buffeting for life the roaring surge;

. . . Often he had snatch'd
From the wild billows, some unhappy man
Who liv'd to bless the hermit of the rocks.[126]

When his rescue efforts were unsuccessful, he would retrieve the mariner's corpse, create a grave for him in the cliff, and then perform a funeral for him:

But if his generous cares were all in vain,
And with slow swell the tide of morning bore
Some blue swol'n cor'se to land; the pale recluse
Dug in the chalk a sepulchre—above
Where the dank sea-wrack mark'd the utmost tide,
And with prayers perform'd the obsequies
For the helpless stranger.[127]

Whether as savior or as eulogist, the hermit offers an example of one whose deep feeling inspires impactful action. Not only is he "feelingly alive," and thus sensitive as well as compassionate, to others and their plights; motivated by his sensitivity, he also serves others, performing abundant "generous cares." By focusing the conclusion of *Beachy Head* on the profound compassion of "the hermit of the rocks," Smith suggests that the practice of sensitive witnessing is beneficial not only because it expands one's philosophical understanding and one's capacity for compassion but also, and perhaps more importantly, because it leads to generous action.

Informing both Smith's speaker's and Hermit Darby's practice of sensitive witnessing is their experience of the self as collective rather than isolated and autonomous. This aspect of Smith's portrayal of sensitive witnessing also closely corresponds to Wollstonecraft's theory of moral sense. We have already observed that Smith emphasizes humanity's entanglement with nonhuman nature, revealing the individual as never truly alone because always mingled with and, thus, touched by and touching nature. The self demanded by Smith's practice of sensitive witnessing is both porous and capacious—open to, and even inclusive of, its environment and those who inhabit it. In fact, I would argue that the calcareous cliff of Beachy Head is its analogue, for the self that Smith envisions is not only, as we have seen, prone to disintegration (via change and ultimately death) but also sedimented (with knowledge and memories) and permeable (to other voices and stories as well as to matter). The poem's opening image of the speaker "reclining" on the cliff thus may be seen as a physical expression of not only their intimacy and continuity but also resemblance. Her use of recollection, intertextuality, and interpolated stories all evoke the porous and capacious quality of the self her practice of sensitive witnessing entails.[128] Such techniques perform the effect that Wollstonecraft argues reason has on emotions. Deepened by reason, they become detached from the individual and become shared or "transpersonal"—no longer feelings of one but, rather, feelings of the many ("humanity").[129] Both Smith and Wollstonecraft perceive the boundaries of the self as provisional. A capacious and collective self is the ultimate goal, for such is the self—the sensitive witness—who not only feels with and for others but also acts in the service of others.

The Gentle Touch of Ecological Sensitivity

As they envision such a sensitive witness, Smith and Wollstonecraft both promote an ontology of kinship and an ethic of care; however, whereas Wollstonecraft focuses on humanity, Smith pursues a more ecological perspective that is inclusive not merely of other humans but of all of earth's creatures.[130] Smith's educational writings explicitly express the importance of protecting the natural world from injury, as is apparent when, in *Minor Morals* (1799), Mrs. Belmour tells her niece Mary, "Take care not to hurt the fly you hold: if you have satisfied your curiosity, let it go; and do not, by robbing its wings of their feathers, deprive the insect of the power of enjoying its little space of life."[131] The ontology of kinship implied by her practice of sensitive witnessing informs this gentleness. One develops a desire to care for nonhuman nature as a result of having perceived oneself to comingle with it. In fact, as the following comment from Mrs. Belmour illustrates, such sensitive perception of comingling can inspire one not only to feel compassion for, but to imagine *becoming*, the nonhuman other: "I have wished myself transformed into a squirrel, that I might live amidst these delightful shades, and bound from bough to bough, finding my food in the beech-nuts, and my shelter among the leaves."[132] In this memory, Mrs. Belmour feels such a sense of kinship with and care for the squirrel that she develops a desire to lose herself entirely so that she might fully consummate the connection between self and other, human and nonhuman nature, that such kinship implies. Furthermore, as the following speech demonstrates, this deep sense of kinship proceeds to inspire an ethic of care that manifests as a determined gentleness:

> MRS. BELMOUR: . . . The active cruelty of man spares not any animal of the creation; and even those which are harmless, and contribute nothing

> to his wants as affording food, are not free from his inhumanity. These poor little inoffensive inhabitants of the woods are objects of chase to idle men and boys, who either shoot them, or throw at them sticks loaded with lead at each end, which generally bring them bleeding and with broken limbs to the ground.
>
> SOPHIA: How barbarous![133]

Though in her adulthood Mrs. Belmour no longer wants to become a squirrel, her sense of kinship with the squirrel and other "animals of creation" persists, inspiring her wish that they would be treated with tenderness. Here, as in many similar passages throughout her schoolbooks and poetry, Smith bemoans humanity's violence against nature.

To return to *Beachy Head*, see this passage in which Smith establishes a contrast between the despicable violence inflicted by "Luxury" and the protection afforded by the shepherd whose sensitive witnessing inspires him to express care through gentle and generous action:

> His boy visits every wired trap
> That scars the turf; and from the pit-falls takes
> The timid migrants, who from distant wilds,
> Warrens, and stone quarries, are destined thus
> To lose their short existence. But unsought
> By Luxury yet, the Shepherd still protects
> The social bird, who from his native haunts
> Of willowy current, or the rushy pool,
> Follows the fleecy croud, and flirt and skims,
> In fellowship among them.[134]

Here, the boy traps the "timid migrants," which Smith's footnote identifies as Wheatears, for "Luxury." Smith here portrays their sacrifice elegiacally, mourning how they must "lose their short existence" to become a culinary delicacy. In such passages, Smith demonstrates that she is aware of the dangers of touch. However, through her tactile poetics, Smith promotes a gentle and sensitive alternative. In the tactile encounters with nature Smith offers in her verse, humans "trace" nature's contours (either with feet or pencils), they "recline" and "repose" on nature's soft surfaces. Even when they take from nature, their touch is light and gentle. The characters that Smith celebrates never "ravage" or "mutilate."[135] In Smith's works, the sensitive witness touches, but they do so carefully and with respect for the other. While the boy "takes . . .

timid migrants" from traps, "the Shepherd . . . protects the social bird"—which Smith's footnote identifies as "The Yellow Wagtail. *Motacilla flava*"—and thus participates in the "fellowship" the wagtail has, as he "flirts and skims," with "the fleecy croud." Mrs. Belmour fantasizes about becoming a squirrel; the shepherd here acts in fellowship with birds. Such is the experience of Smith's sensitive witness—they perceive, feel, identify, and care for those humans and nonhumans who are in need of care.

• • •

That melancholic sensibility features prominently in Smith's writing is unsurprising. Born into wealth and privilege but then subjected to poverty, trauma, and loss as an adult, she certainly had plenty to mourn. Her difficulties began when her father married her off at the age of fifteen to Benjamin Smith who proved incapable of sustaining his growing family. Due to her husband's incompetence and recklessness, she suffered from many years of severe financial hardship and even spent several months in debtors' prison at her husband's side.[136] Moreover, she became embroiled in a protracted legal battle, brought on by complications related to her father-in-law's will. This legal battle became legendary and even appears to have inspired Charles Dickens's depiction of the Jarndyce and Jarndyce suit in *Bleak House*.[137] As Sarah M. Zimmerman relates,

> Wanting to protect his legacy of approximately £36,000 from his unreliable son and to secure it for his grandchildren, [Richard Smith] undermined his own efforts, however, in an intricate will that proved vulnerable to legal challenges. After his death in 1776 the trustees could not agree on its meaning, and a chancery suit was spawned that did not reach even a partial settlement until 1798, and remained open until after Charlotte Smith's death. The estate's anticipated settlement prompted Smith's writing career: she wrote in order to maintain her children's social standing until they received their inheritance.[138]

Smith gave birth to a total of twelve children, but only six of them survived her, and the deaths of her children were possibly the greatest hardships of her life. "No disadvantage could equal those I sustained," Smith confided to a friend. "[T]he more my mind expanded, the more I became sensible of personal slavery."[139]

However, Smith's melancholic sensibility is not simply a result of her own personal traumas, nor is it an affectation or a technique for cultivating literary popularity by appealing to the cult of sensibility. Rather, it is indicative of Smith's evolving commitment to a philosophical practice of sensitive witnessing that she develops and promotes for the sake of empirical discovery, aesthetic pleasure, and compassionate social reform. Though traces of this practice may be found throughout several of her works, it reaches its fullest expression in the final poem of her career, *Beachy Head*. It is there that Smith attests to its philosophical as well as its ethical value, maintaining its superiority to more conventional modest witnessing because of its capacity both to reveal empirical truths and to inspire generous action. Informed by sensitive witnessing, Smith's distinctive variety of empiricism entails a mixed method of poetry and science, an ontology of kinship, and an ethic of care. Moreover, it harnesses not only careful observation but also feeling and reason and, in doing so, maximizes its insight and benefit to the public good. Thus, Smith channels the popularity and power of sensibility for the sake of poetry and philosophy while distinguishing it from more conventional expressions of sensibility that tended to result in self-indulgence as well as cliché. The sensitive witness is, for Smith, not simply of value to the collective pursuit of knowledge but also to social justice and the greater good.

CODA

IN THEIR 2008 BOOK *Material Feminisms*, Stacy Alaimo and Susan J. Hekman posit and seek to help foster a "'material turn' in feminist theory, a wave of feminist theory that is taking matter seriously."[1] As Alaimo and Hekman acknowledge, this turn coincides with the advent of feminist "new materialism," which also emerged in the early twenty-first century to encourage a newly invigorated reckoning with matter in the face of poststructuralism's privileging of discourse. *Sensitive Witnesses* shows that such feminist materialism has a deep history. Each of the female philosophers we have here explored share with feminist new materialists the goals of taking matter seriously and of finding "a way to talk about the materiality of the body as itself an active, sometimes recalcitrant, force."[2] They also similarly push back against the "nature/culture dualism," denying an "epistemology [that] is grounded in objective access to a real/natural world" while at the same time nevertheless seeking to access nature as they bear witness to their own materiality.[3]

In her influential book *Vibrant Matter: A Political Ecology of Things*, which develops an ecologically oriented theory and practice of "vital materialism," Jane Bennett calls for a "newish self" constituted by "the oxymoronic truism that the human is not exclusively human, that we are made up of its."[4] Bennett recognizes that there are significant, because deeply rooted and felt, barriers to cultivating such a self:

> Even if, I as [*sic*] believe, the vitality of matter is real, it will be hard to discern it, and, once discerned, hard to keep focused on. It is too close and too fugitive, as much wind as thing, impetus as entity, a movement

> always on the way to becoming otherwise, an effluence that is vital and engaged in trajectories but not necessarily intentions. What is more, my attention will regularly be drawn away from it by deep cultural attachments to the ideas that matter is inanimate and that real agency belongs only to humans or to God, and by the need for an action-oriented perception that must overlook much of the swirling vitality of the world.[5]

However, Bennett insists that so important is the cultivation of a self informed by the vitality of matter that we me must be determined to take on such challenges. Not only is vital materialism more accurate than its conventional alternative, which wrongly perceives matter as humanity's inert opposite. It is also more equipped to preserve ecological health, which is of course an increasingly pressing need as we have only just begun to witness the devastating effects of our centuries-old compulsion "to produce and consume in . . . violently reckless ways."[6] Vital materialism is, for Bennett, a "truism," and the sooner we acknowledge it as such, the better:

> Admit that humans have crawled or secreted themselves into every corner of the environment; admit that the environment is actually inside human bodies and minds, and then proceed politically, technologically, scientifically, in everyday life, with careful forbearance, as you might with unruly relatives to whom you are inextricably bound and with whom you will engage over a lifetime, like it or not. Give up the futile attempt to disentangle the human from the nonhuman. Seek instead to engage more civilly, strategically, and subtly with the nonhumans in the assemblages in which you, too, participate.[7]

We must, according to Bennett, "start to feel [ourselves] as not only human. . . . [A]n affective, speaking human body is not *radically* different from the affective, signaling nonhumans with which it coexists, hosts, enjoys, serves, consumes, produces, and competes."[8]

But how? How can we experience ourselves as "made up of its" when there is so much preventing us from doing so?[9] The "buffering" that Charles Taylor attributes to the modern self finds its apotheosis in the masculine modesty that Bacon, Burke, and Kant presented as necessary for empirical and aesthetic philosophical practice.[10] Even though their philosophical observations and principles often challenged such modesty (as in the case of Boyle's corpuscularianism), British Enlightenment philosophers nevertheless regularly performed, and promoted the perception of, a fundamental distinction between active man and inert matter. The vital materialism that Bennett promotes challenges precisely such a binaristic distinction—such a buffering—instead registering "the force of things" or what she calls "thing power." According to Bennett, materiality is, itself, a "creative agent."[11] Inspired by Samuel Taylor Coleridge, Bennett argues that such "universal creativity requires a special sensitivity."[12] It is precisely such "sensitivity" that the author-philosophers we have here explored model and promote. To be a sensitive witness is to recognize matter as an agent and also to appreciate one's own entanglement—as well as one's continuity—with the material world. The polemic that Bennett presents expresses sentiments that several prominent female author-philosophers also expressed in the Enlightenment era. *Sensitive Witnesses* has sought to recover their reactions and contributions to "Enlightenment subjectivity" as a means of recognizing that, in the Enlightenment era, there were those who insisted on bearing witness to the vitality of matter and the continuity of humanity with nonhuman nature—who rejected the binarisms that helped substantiate the modern self's buffering.

Masculine philosophical modesty is a fiction created and perpetuated to preserve the "purity" of empirical and aesthetic experience. However, in spite of philosophers' attempts to normalize and universalize it, we must recognize masculine modesty as a historical development and a privileged experience. The truth of the matter is that our self is itself a fiction, particularly if defined as a solitary individual, separate from the world around it. Whether we like it or not, we are always touching nature, though it takes a certain bravery and resilience to acknowledge it. Touching nature is, after all, sublime. But even as such, it flies in the face of tradition because the sublime it entails is dangerous, physical, highly palpable. Rather than keeping that which is overwhelming at a remove, it invites the astonishing, overwhelming, annihilating in. Touching nature, the sensitive witness communes with, rather than simply rationalizing, the sublime.

In chapter 4, we conclude with Charlotte Smith's ethic of care. Such an ethic has been expressed as well as complicated by recent feminists.[13] I would argue that neither Smith nor the other authors that *Sensitive Witnesses* studies are essentialist and, therefore, that the ethic of care that Smith advocates, inspired by her practice of sensitive witnessing, is as important for men as it is for women. In fact, she recommends such an ethic to girls and boys alike in her works for children. The author-philosophers that *Sensitive Witnesses* studies seek to harness characteristics that have become associated with femininity—

sensitivity, vulnerability, openness—but they do so with the hope of encouraging all, both women and men, to adopt such characteristics. Their argument is that such characteristics will promote better philosophy. However, for Charlotte Smith, the final author-philosopher *Sensitive Witnesses* studies, they will also promote a healthier public good. The ethic of care should not simply rest on the shoulders of women, according to Smith. Rather, it should be shared by all who are in a position to care for others. Equipped with an ontology of kinship, the ethic of care knows no bounds.

In 1818, Mary Shelley anonymously published the first edition of *Frankenstein*. I would argue that the feminist philosophical practices *Sensitive Witnesses* studies set an important foundation for her "hideous progeny."[14] In *Frankenstein*, Shelley directly challenges the modest witness with her portrayal of Victor, "the modern Prometheus," who is so self-involved—so buffered—that he has absolutely no sense of how his actions and choices will impact others. Victor is oblivious, and in portraying him as such, Mary Shelley challenges the obliviousness of modest witnessing more generally. More attention needs to be given to how Mary Shelley's masterpiece is reminiscent of earlier works by Enlightenment female authors. Though Clerval and Elizabeth are, in certain respects, evocative of the Enlightenment sensitive witness in their embodiment of poetic sensibility, neither of these characters is a philosopher per se. The Creature himself perhaps comes closest to the Enlightenment era sensitive witness in the novel's central story, particularly in his pursuit of education through boundary-defying empirical practice (as is evinced by his painful "examination" of fire), the emotional transport ("reverence") he experiences as he encounters the De Laceys, and his harnessing of the sublime (most spectacularly, when he sends the De Laceys' cottage up in flames).[15] However, the one true philosopher central to *Frankenstein*—namely, Victor—determinedly refuses to embody a sensitive witness, as is illustrated in his initial responses to the Creature's "birth":

> "Why do you call to my remembrance circumstances of which I shudder to reflect, that I have been the miserable origin and author? Cursed be the day, abhorred devil, in which you first saw light! Cursed (although I curse myself) be the hands that formed you! You have made me wretched beyond expression. You have left me no power to consider whether I am just to you, or not. Begone! Relieve me from the sight of your detested form."

Here, as he refuses to feel for or with the Creature, Victor Frankenstein seeks to distance himself from his own creation and, thus, I would argue to purify himself of matter by, as was typical within British Enlightenment science, denying his own entanglement with it. The remainder of the novel dramatizes the tragic consequences of such a masculine and modest refusal of sensitivity. Moreover, through the incessant doubling for which the novel is known, Shelley criticizes Victor's refusal and the purification it attempts to achieve as delusional. Try as he might to deny it, Victor cannot help but be entangled with the Creature. After all, he is himself a material being with a touching nature. That Shelley emphasizes the abject materiality of Victor's act of creation as well as its tactile intimacy—"Cursed . . . be the hands that formed you!"—only accentuates Victor's entanglement with matter. So too does the doppelgänger relationship that Shelley establishes between Victor and the Creature. If Victor had chosen to recognize his own kinship with the Creature and thus been inspired to care for him rather than to abandon him, it is likely that the novel's many tragedies would have been avoided. Of course, in the end, Victor's desire to achieve immortality is revealed to be not only tragic but also profoundly ironic because it perpetuates death rather than life.

Frankenstein is regularly analyzed as a critique of modern science, and various scholars have pursued this analysis through an ecocritical interpretive framework.[16] By studying the Enlightenment history of the sensitive witness, we acquire new ways of recognizing the meaning and significance of both the novel's critique of science and its ecological dimensions and legacy. I would suggest that, contextualized by the history of the sensitive witness, *Frankenstein* becomes an elegy for philosophical possibilities that, by the time of the novel's publication in 1818, appear to have been foreclosed upon. The march of progress—scientific and philosophical as well as sociopolitical and economic—has not been conducive to sensitive witnessing. *Frankenstein* portrays a confrontation between sensitive witnessing and modest witnessing in which, ultimately, the latter wins out. As it does so, the novel may be seen to narrate the historical success of patriarchal modernity, and the related ascendance of the buffered modern self, as a multi-pronged tragedy in which everyone seems destined to suffer.

If Shelley leaves us with any hope, it resides in Robert Walton and his "dear sister," Margaret. In the novel's epistolary conclusion, Walton tells his sister of two occasions in which he has been a sensitive witness. The first involves his bearing witness to Victor's telling of his story, which culminates in his death. Walton responds to Victor's story and death thus: "Margaret, what comment can I make on the untimely extinction of this glorious spirit? What can I say, that will enable you to understand the depth of my sorrow? All that I should express would be inadequate and feeble. My tears flow; my mind is overshadowed by a cloud of disappointment. But I journey towards England, and I may there find consolation."[17] The second occasion of Walton's sensitive witnessing involves the Creature who has found his way into "the cabin where the remains of Frankenstein still lie."[18] Walton explains to Margaret that, instead of being repelled by the Creature as others throughout the novel have been, and in spite of his initial "impulses" to fulfill Victor's "dying request" that he kill the Creature, he is "suspended by a mixture of curiosity and compassion."[19] He then proceeds to "approach . . . [the] tremendous being" and to listen to him tell his own story. When the Creature comes to a stopping point, Walton explains, "I was at first *touched* by the expressions of his misery."[20] Here, I would argue, Walton reveals himself to be a sensitive witness as he does not simply listen to the Creature with sympathy at a distance but, rather, allows himself to be "touched" by his "expressions." Indeed, after so much alienation throughout the novel, the intimacy that immediately exists between Walton and the Creature is remarkable.

Even when Walton adjusts his initial reaction to the Creature, he nevertheless continues to listen carefully to what he has to say and to document it, as if word for word, for his sister. He acknowledges to Margaret, "when I called to mind what Frankenstein had said of [the Creature's] powers of eloquence and persuasion, and when I again cast my eyes on the lifeless form of my friend, indignation was re-kindled within me."[21] Walton and the Creature then proceed to engage in a brief dialogue about the latter's "misery," but almost immediately Walton allows the Creature to continue his monologue without any additional commentary.[22] In fact, Walton and Shelley allow the Creature's story to conclude the novel, which has the effect of reinforcing both Walton's and Shelley's status as sensitive witnesses. In the final lines of the novel, the Creature envisions his own funeral pyre, imagining his "ashes [being] swept into the sea by the winds" and his "spirit . . . sleep[ing] in peace."[23] All that follows are two baldly descriptive sentences by Walton: "He sprung from the cabin-window, as he said this, upon the ice-raft which lay close to the vessel. He was soon borne away by the waves, and lost in darkness and distance."[24] Thus, though he has been demonized throughout the novel, and has repeatedly been a vehicle for violence and death, the Creature is nevertheless given the last word.

Where does this leave the novel's reader? I would contend that, ultimately, the novel invites the reader to occupy a position parallel to Walton's and

Shelley's and to emulate their sensitivity as witnesses. Shelley encourages the reader to be, like Walton, "touched" by the tragic story while also being judicious in how they evaluate the story's nuances and details. Like her mother, Mary Wollstonecraft, as well as Charlotte Smith, Mary Shelley promotes a combination of sensitivity (or sensibility) and reason whereby the latter both "deepens" and provides a check on the former.[25] Like her mother, Shelley is suspicious of "mechanical instinctive sensations."[26] It is not enough, according to Shelley, to be "touched," as Walton initially is when he first encounters the Creature. With the conclusion of *Frankenstein*, she encourages her readers to proceed, like Walton does, to engage in rational reflection so that they may learn how to revise their initial "mechanical," "instinctive," and purely emotional responses such that these responses become "deepened" in the terms Wollstonecraft outlines—more nuanced and more responsive to the broader picture (including *both* the Creature's "expressions of his misery" *and* "what Frankenstein had said").[27] Walton thus ultimately serves as a sensitive witness who deserves to be emulated.

By casting Walton as a sensitive witness, Mary Shelley takes a strong anti-essentialist position, explicitly recognizing that there is nothing *natural* about the binaristic separation between masculine modesty and feminine sensitivity. If a brave explorer like Walton can be a sensitive witness, surely it is possible for any man. In fact, I would argue that Walton and Margaret's relationship serves as a foil to Victor and Elizabeth's. The latter expresses the patriarchal polarization of modest men and sensitive women, which found expression in what has become known as "the separation of the spheres," while the former represents Shelley's progressive hope for a more integrated, less binaristic, and more egalitarian relationship between men and women.[28] As she juxtaposes these relationships, she also juxtaposes Victor and Walton. In so doing, she implicitly calls for a revision to scientific practice that entails a shift away from modesty to sensitivity—or, in the terms of Haraway, a "queering" of the modest witness.[29] We have already recognized how *Frankenstein* challenges the ethics as well as the authority of the masculine modest witness, as embodied by Victor. I would argue that, with her favorable portrayal of Walton, particularly at the conclusion of the novel, she presents a strong endorsement of the ethics as well as the authority of the sensitive witness while also insisting that men may—and should—embody it as well as women.

Ultimately, then, though the main story of *Frankenstein* functions as an elegy for philosophical possibilities foreclosed upon, with her portrayal of the peripheral characters of Walton and Margaret, Shelley nevertheless holds onto hope for a different future. The future she invites her readers to help envision is one in which science and discovery will be guided not by masculine modesty that entails an alienation of man from matter but, rather, by an androgynous sensitivity that combines feeling and reason while acknowledging instead of denying humanity's entanglement with the material world. Such is the future that the author-philosophers throughout *Sensitive Witnesses* sought, and we have never needed such a future as much as we do now. As we have begun to experience, firsthand, the devastating effects of anthropogenic ecological degradation such as devastatingly rampant forest fires, flooding that displaces large populations, disrupted weather patterns that put many people in harm's way, etc., the fiction of modest man's ability to separate himself from matter is becoming increasingly difficult to maintain as truth. Few have recognized this difficulty more profoundly than climate scientists. As a recent study by Lesley Head and Theresa Harada shows, "emotional management is an important part of dealing with climate change knowledge in both the professional and the domestic spheres" of climate scientists.[30] Though the climate scientists in their study continue to strive to "compl[y] with the scientific norms of the doing of climate science . . .—which requires an objective standpoint," the act of repressing their emotions, and thus denying their sensitivity, requires significant emotional labor, which threatens not only to hinder their scientific work but also to perpetuate the relative lack of diversity amongst climate scientists.[31] Moreover, within climate science, emotion can have salutary effects. As Head and Harada's study demonstrates, "emotions associated with the challenges of climate science helped to inspire and motivate scientists to 'keep on going.' . . . [T]his fortitude to persevere was one of the ways that emotions worked to bind them together."[32] Head and Harada conclude their study with the following claim: "Discussing how we feel about climate change will at the very least broaden our thinking and give us more options for effective action."[33] Surely, this is true not only for lay people but also for scientists themselves.

I would argue that the time has come for us to adjust our scientific methods to recognize, along with the several pioneering Enlightenment era author-philosophers we have here explored, that emotion and empiricism need not be mutually exclusive and, in fact, may even be mutually supportive. As climate catastrophes remind us of the undeniability of our entanglement with matter, we simply no longer have the luxury of achieving objectivity by "purifying" ourselves of matter. To combat climate change and the environmental injustices that it has already begun to inflict, we would all do well to develop a practice of sensitive witnessing that is supportive of an ontology of kinship and ethic of care, as is the case in the final work that *Sensitive Witnesses* explores—Charlotte Smith's *Beachy Head*. In the words of Robin Wall Kimmerer—moss scientist, member of the Citizen Potawatomi Nation, and twenty-first-century sensitive witness—"[t]he story" of our material world is "about relationship."[34] How different our present and our future would be if we were all to "come to see the world" not at a distance but, like Kimmerer does, "through moss-colored glasses."[35]

Cary Wolfe has argued that posthumanism "comes both before and after humanism."[36] On the one hand, it is a "historical moment," marking a shift away from humanism, "in which the decentering of the human by its imbrication in technical, medical, informatic, and economic networks is increasingly impossible to ignore."[37] However, less obviously, it also *precedes* humanism because, according to Wolfe, "it names the embodiment and embeddedness of the human being in not just its biological but also its technological world."[38] I argue that the Enlightenment era sensitive witness evokes this pre-humanist dimension of posthumanism. As it does so, it offers a historical precursor to forms of witnessing and ethical practices that ecologically minded posthumanist philosophers like Wolfe have developed in recent decades. We therefore have much to learn from Enlightenment era sensitive witnesses—from their accomplishments as well as the challenges they faced. A central lesson is that, if we are to learn how to adapt sensitive witnessing to the twenty-first century so that we may become better prepared to face and address the devastating ecological consequences of centuries of modest witnessing, we must develop a new, more capacious perception of ourselves that includes rather than excludes others, both human and nonhuman. Furthermore, we must accept, and even learn how to appreciate, knowledge-making as a sublime endeavor, though one that is predicated not on distance, but rather on physical and emotional intimacy. So profound is such intimacy that we will be required to lose ourselves (including our egos, expectations, ambitions, etc.) as we seek to know and to act on our knowledge. We can no longer afford the arrogance of modesty. It is time for a new, sensitive, "great instauration" in which we learn how to touch and be touched by nature, with a sense of continuity, relationality, gentleness, and care.

NOTES

Introduction

1. On the context and reception of Swift's *Modest Proposal*, see especially Ian McBride, "The Politics of *A Modest Proposal*: Swift and the Irish Crisis of the Late 1720s," *Past & Present* 244, no. 1 (2019): 89–122; Charlotte Sussman, "The Colonial Afterlife of Political Arithmetic: Swift, Demography, and Mobile Populations," *Cultural Critique* 56 (2004): 96–126.

2. "Modest," *Oxford English Dictionary Online*.

3. Ibid.

4. Donna J. Haraway, *Modest_Witness@Second_Millenium.FemaleMan_Meets_OncoMouse: Feminism and Technoscience* (New York: Routledge, 1997), 30.

5. Sussman, "The Colonial Afterlife of Political Arithmetic," 96.

6. Jonathan Swift, *A Modest Proposal, The Writings of Jonathan Swift*, ed. Robert A. Greenberg and William B. Piper (New York: Norton, 1973), 504.

7. Steven Shapin and Simon Schaffer, *Leviathan and the Air-Pump: Hobbes, Boyle, and the Experimental Life* (Princeton, NJ: Princeton University Press, 1985), 65. The phrase "naked way of writing" is from Robert Boyle, "Proëmical Essay . . . with some Considerations touching Experimental Essays in General," in *The Works of the Honourable Robert Boyle*, vol. 1, 2nd ed., ed. Thomas Birch (London: J. & F. Rivington, 1772), 318.

8. *The Record of the Royal Society of London for the Promotion of Natural Knowledge*, 4th ed. (London: Morrison & Gibb, 1940), 290.

9. Tita Chico, *The Experimental Imagination: Literary Knowledge and Science in the British Enlightenment* (Stanford, CA: Stanford University Press, 2018), 26.

10. On the evolution of Baconian modesty into objectivity, see Lorraine Daston and Peter Galison, *Objectivity* (Brooklyn: Zone Books, 2007). The phrase "the view from nowhere" is from Thomas Nagel, *The View from Nowhere* (Oxford: Oxford University Press, 1989).

11. National Academy of Sciences, National Academy of Engineering, and Institute

of Medicine, *Responsible Science: Ensuring the Integrity of the Research Process: Volume I* (Washington, DC: The National Academies Press, 1992), 38.

12. Francis Bacon, *Novum Organum*, in *The Works of Francis Bacon*, ed. James Spedding, Robert Leslie Ellis, and Douglas Denon Heath, vol. 8 (New York: Hurd and Houghton, 1864), 76–77, 99.

13. My emphasis; qtd. in G. J. Barker-Benfield, *The Culture of Sensibility: Sex and Society in Eighteenth-Century Britain* (Chicago: University of Chicago Press, 1996), 23.

14. "Delicate," *Oxford English Dictionary Online*. See also "womanish."

15. Robert Whytt, *Observations on the Nature, Causes and Cure of those Disorders which have been Commonly called Nervous, Hypochondriac or Hysteric* (Edinburgh, 1765), 115.

16. On the perceived sensitivity of women's nerves, see esp. Barker-Benfield, *The Culture of Sensibility*; G. Rousseau, *Nervous Acts: Essays on Literature, Culture and Sensibility* (Hampshire: Palgrave Macmillan, 2004). For a helpful overview of the gendering of nerves in the British Enlightenment, see also Eleanor C. L. Crouch, "Nerve Theory and Sensibility: 'Delicacy' in the Work of Fanny Burney," *Literature Compass* 11, no. 3 (2014): 206–17.

17. Charles Taylor, *A Secular Age* (Cambridge: Cambridge University Press, 2007), 37–38.

18. John Locke, *Second Treatise of Government*, ed. C. B. Macpherson (Indianapolis, IN: Hackett Publishing Co., 1980), 19.

19. See especially, Jennifer Rae Greeson, "The Prehistory of Possessive Individualism," *PMLA* 127, no. 4 (2012): 918–24; Charles Taylor, "Locke's Punctual Self," in *Sources of the Self: The Making of the Modern* Identity (Cambridge, MA: Harvard University Press, 1989), 159–76.

20. Locke, *Second Treatise*, 19.

21. Ibid., 21.

22. Ruth Perry, "Mary Astell and the Feminist Critique of Possessive Individualism," *Eighteenth-Century Studies* 23, no. 4 (1990): 452.

23. Mary Astell, *Some Reflections upon Marriage Occasion'd by the Duke & Dutchess of Mazarine's Case, Which Is Also Considered.* (London: Printed for John Nutt, 1700), 72.

24. Leslie Richardson, "Leaving Her Father's House: Astell, Locke, and Clarissa's Body Politic," *Studies in Eighteenth-Century Culture* 34 (2005): 152.

25. I appreciate Kathleen Lubey's suggestion that the use of "'female' and 'male' as adjectives . . . risk[s] positing an anatomical sexual binary" and "assert[ing] causality between genitals and gender." Like Lubey, I oppose binaristic and/or essentialist understandings of gender. In fact, the authors whom I address throughout *Sensitive Witnesses* regularly do so as well. However, I have nevertheless chosen to use "female" and "male" as adjectives throughout the book for one main reason: these were terms that were regularly used in Enlightenment Britain as adjectives to signify gender and are thus historically precise. Additionally, as a historian and across many contexts, I have found them more neutral and therefore versatile than the terms "woman" and "man."

Kathleen Lubey, *What Pornography Knows: Sex and Social Protest Since the Eighteenth Century* (Stanford, CA: Stanford University Press, 2022), 27.

26. Karen Gevirtz has also explored how Enlightenment era female authors pose significant challenges to masculine modest witnessing. However, whereas Gevirtz's study focuses almost solely on narrative fiction, and is primarily interested in the literary consequences of authors' philosophical practices and principles, *Sensitive Witnesses* studies a broad variety of genres and attends equally to their philosophical and literary consequences. Furthermore, Gevirtz neglects to consider the influence of materialism on Enlightenment era female authors. In recognizing this influence, the present study encourages a fuller and more nuanced appreciation of British women's claim to philosophical authority in the Enlightenment period. Karen Bloom Gevirtz, *Women, the Novel, and Natural Philosophy, 1660–1727* (New York: Palgrave Macmillan, 2014).

27. The term "trans-corporeality" comes from Stacy Alaimo's highly influential book *Bodily Natures.* With this concept, she seeks to encapsulate, and to provide a framework for theorizing, the way in which "the human is always intermeshed with the more-than-human world" such that "the substance of the human is ultimately inseparable from 'the environment.'" Alaimo uses the term "trans-corporeality" to "think . . . across bodies" so as to "catalyze the recognition that the environment, which is often imagined as inert, empty space or as a resource for human use, is, in fact, a world of fleshy beings with their own needs, claims, and actions. By emphasizing the movement across bodies, trans-corporeality reveals the interchanges and interconnections between various bodily natures." Stacy Alaimo, *Bodily Natures: Science, Environment, and the Material Self* (Bloomington: Indiana University Press, 2010), 2, 4. The term "entanglement" comes from Donna Haraway, "Anthropocene, Capitalocene, Plantationocene, Chthulucene: Making Kin," *Environmental Humanities* 6 (2015): 160.

28. My concept of "material feminism" is especially informed by Stacy Alaimo and Susan J. Hekman, *Material Feminisms* (Bloomington: Indiana University Press, 2008). See also Sara Ahmed, "Imaginary Prohibitions: Some Preliminary Gestures on the Founding Gestures of the 'New Materialism,'" *European Journal of Women's Studies* 15, no. 1 (2008): 23–39; Karen Barad, *Meeting the Universe Halfway: Quantum Physics and the Entanglement of Matter and Meaning* (Durham, NC: Duke University Press, 2007); Diana Coole and Samantha Frost, *New Materialisms: Ontology, Agency, Politics* (Durham, NC: Duke University Press, 2010); Susan Hekman, *The Material Knowledge: Feminist Disclosures* (Bloomington: Indiana University Press, 2010).

29. Chico, *Experimental Imagination*; Helen Thompson, *Fictional Matter: Empiricism, Corpuscles, and the Novel* (Philadelphia: University of Pennsylvania Press, 2017); Al Coppola, *The Theater of Experiment: Staging Natural Philosophy in Eighteenth-Century Britain* (Oxford: Oxford University Press, 2016); Alexander Wragge-Morley, *Aesthetic Science: Representing Nature in the Royal Society of London, 1650–1720* (Chicago: University of Chicago Press, 2020); Peter Hanns Reill, *Vitalizing Nature in the Enlightenment* (Berkeley: University of California Press, 2005). See also Lorraine Das-

ton, "On Scientific Observation," *Isis* 99 (2008): 97–110; Jessica Riskin, *Science in the Age of Sensibility: The Sentimental Empiricists of the French Enlightenment* (Chicago: University of Chicago Press, 2002); Charles T. Wolfe and Ofer Gal, eds., *The Body as Object and Instrument of Knowledge: Embodied Empiricism in Early Modern Science* (Dordrecht, Netherlands: Springer, 2010).

30. Abraham Cowley, "To the Royal Society," in *The History of the Royal-Society of London for the Improving of Natural Knowledge*, by Thomas Sprat (London, 1667), n.p; Chico, *Experimental Imagination*, 1.

31. Wragge-Morley, *Aesthetic Science*, 3.

32. Bacon, *Novum Organum*, 5.

33. Thompson, *Fictional Matter*, 3. Other prominent descriptions of Enlightenment era modest witnessing Thompson similarly challenges are Barad, *Meeting the Universe*; Michael McKeon, *The Origins of the English Novel, 1600–1740* (Baltimore: Johns Hopkins University Press, 1987).

34. Thompson, *Fictional Matter*, 33.

35. Ibid., 34.

36. Ibid., 279. This presumptive view of the Enlightenment era has become ubiquitous in twenty-first century environmental humanities. For a particularly multifaceted recent representation of it, see the special issue of *Telos* devoted to the Anthropocene in 2016 wherein, for example, Timothy Sean Quinn finds in Francis Bacon's philosophy "the beginnings of the modern worldview: a duality of a cosmos without purpose or intrinsic meaning, and a purposive world of human action upon that indifferent world, developing technologies that aid us in dominating it." Timothy Sean Quinn, "Anthropocene Crises and the Origin of Modernity," *Telos* 177 (winter 2016): 47. Of course, as foundational works by Horkheimer and Adorno as well as Carolyn Merchant attest, such an identification between early modern scientific innovations and the environmental exploitation that would come to distinguish later centuries is far from new. Max Horkheimer and Theodor W Adorno, *Dialectic of Enlightenment* (1944), trans. John Cumming (New York: Continuum, 1982); Carolyn Merchant, *The Death of Nature: Women, Ecology, and the Scientific Revolution* (New York: Harper & Row, 1980).

37. Thompson, *Fictional Matter*, 15. For example, Principe provides evidence that Newton suffered from "mercury poisoning." Lawrence M. Principe, *The Aspiring Adept: Robert Boyle and His Alchemical Quest* (Princeton, NJ: Princeton University Press, 1998), 179. A particularly illustrative example of the intense physicality of philosophers' engagement with their specimens is Robert Boyle's *Workdiaries*. Robert Boyle, *The Workdiaries of Robert Boyle* (1647–1691), eds. Michael Hunter and Alison Wiggins, CELL website, Centre for Editing Lives and Letters, 2001, http://www.livesandletters.ac.uk/wd/.

38. See for instance Thompson, *Fictional Matter*, 66–107; Peter R. Anstey, *John Locke and Natural Philosophy* (Oxford: Oxford University Press, 2011); Stephen Gaukroger, "The Role of Natural Philosophy in the Development of Locke's Empiricism," *British Journal for the History of Philosophy* 17, no. 1 (2009): 55–83; Michael Hunter, *Boyle:*

Between God and Science (New Haven, CT: Yale University Press, 2009); J. R. Milton, "Locke at Oxford," in *Locke's Philosophy: Content and Context*, ed. G. A. J. Rogers (Oxford: Clarendon Press, 1994), 29–48; J. R. Milton, "Locke's Life and Times," in *The Cambridge Companion to Locke*, ed. Vere Chappell (Cambridge: Cambridge University Press, 1994), 5–25; G. A. J. Rogers, "The Intellectual Setting and Aims of the Essay," in *The Cambridge Companion to Locke's* Essay Concerning Human Understanding, ed. Lex Newman (Cambridge: Cambridge University Press, 2007), 7–32; Patrick Romanell, *John Locke and Medicine: A New Key to Locke* (Buffalo, NY: Prometheus Books, 1984); Rose-Mary Sargent, *The Diffident Naturalist: Robert Boyle and the Philosophy of Experiment* (Chicago: University of Chicago Press, 1995); John W. Yolton, *Locke and the Compass of Human Understanding: A Selective Commentary on the 'Essay'* (Cambridge: Cambridge University Press, 1970).

39. Thompson, *Fictional Matter*, 106.

40. Ibid., 79.

41. Ibid., 79.

42. Ibid., 80. As Thompson recognizes, "[t]he location of Locke's secondary qualities marks a sustained source of philosophical dispute," hence the controversial nature of her argument. Thompson, *Fictional Matter*, 296n37.

43. Ibid., 80.

44. Jay L. Garfield, *The Concealed Influence of Custom: Hume's* Treatise *from the Inside Out* (Oxford: Oxford University Press, 2019), 123–24.

45. Ibid., 124–25.

46. Ibid., 127.

47. Ibid., 173–78.

48. My emphasis; David Hume, *A Treatise of Human Nature* (1739), ed. Ernest C. Mossner (London: Penguin Books, 1969), 234.

49. Garfield, *Concealed Influence*, 154.

50. Ibid., 93–4. In fact, according to Garfield, Hume's "social account of the passions . . . is [his] greatest contribution to the moral psychology of the eighteenth century." It is "the heart of the Humean project." Garfield, *Concealed Influence*, 58.

51. Ibid., 98–99.

52. My emphasis; ibid., 99–101.

53. Ibid., 95n6.

54. Thompson, *Fictional Matter*, 2.

55. Haraway, *Modest_Witness@Second_Millenium*, 32, 24.

56. "Witness," *Oxford English Dictionary Online*.

57. Robert Hooke, *Micrographia* (London, 1665), 171.

58. In an article on what I call "tactile microscopy," I read the detachment of this passage by Hooke in relationship to the satires of science presented by Jonathan Swift's "Voyage to Brobdingnag" in *Gulliver's Travels* (1726) and by an anonymous pornographic poem entitled "The Microscope" (1732). Kristin M. Girten, "Mingling with Matter: Tactile Microscopy and the Philosophic Mind in Brobdingnag and Beyond," *The Eighteenth Century*, 54, no. 4 (Winter 2013): 508–9.

59. Coppola, *Theater of Experiment*, 173.

60. Ibid., 175.

61. Ibid., 174.

62. Joseph Priestley, preface to *The History and Present State of Electricity, with Original Experiments* (London, 1769), i.

63. Ibid., ii.

64. Priestley, *History*, 550.

65. Coppola, *Theater of Experiment*, 174.

66. In *Science in the Age of Sensibility*, Jessica Riskin makes a strong case for what she calls "sentimental empiricism" in the context of the French Enlightenment, and Coppola relies heavily on her study. However, it is important not to assume that British empiricists were inclined to invest in sentiment and sensibility as their French counterparts were. More research needs to be conducted to develop a precise understanding of how British empirical philosophers of the Enlightenment era engaged (or resisted engagement with) sentiment and sensibility.

67. Caroline Herschel, "An Account of a New Comet. In a Letter from Miss Caroline Herschel to Charles Blagden, M. D. Sec. R. S.," *Philosophical Transactions of the Royal Society of London* 77 (1787): 1.

68. William Herschel, "Remarks on the New Comet. In a Letter from William Herschel, LLD. F. R. S. to Charles Blagden, M. D. Sec. R. S.," *Philosophical Transactions of the Royal Society of London* 77 (1787): 4–5. For a fuller analysis of the relationship between Caroline's account and William's subsequent remarks, see Emily Winterburn, "Learned Modesty and the First Lady's Comet: A Commentary on Caroline Herschel (1787) 'An account of a new comet,'" *Philosophical Transactions: Mathematical, Physical and Engineering Sciences* 373, no. 2039 (April 13, 2015): 1–11.

69. William Herschel, "Remarks," 2.

70. Longinus, *On the Sublime*, trans. William Smith (Dublin, 1740), 11.

71. My emphasis; Edmund Burke, *A Philosophical Enquiry into the Origin of Our Ideas of the Sublime and Beautiful* (1757), ed. James T. Boulton (Notre Dame, IN: University of Notre Dame Press, 1968), 68.

72. See for example Paul Crutzen and Eugene Stoermer, "The 'Anthropocene,'" *IGBP Newsletter* 41 (2000): 17–18; Timothy Morton, *Hyperobjects: Philosophy and Ecology after the End of the World* (Minneapolis: University of Minnesota Press, 2013), 7.

73. Aaron Hanlon, "Margaret Cavendish's Anthropocene Worlds," *New Literary History* 47 (2016): 65.

74. Lynn Festa, *Fiction without Humanity: Person, Animal, Thing in Early Enlightenment Literature and Culture* (Philadelphia: University of Pennsylvania Press, 2019), 39.

75. Amanda Jo Goldstein, *Sweet Science: Romantic Materialism and the New Logics of Life* (Chicago: University of Chicago Press, 2017), 18.

76. Ibid., 7, 18.

77. Ibid., 8.

78. For examples of studies documenting this influence, see for instance Annette

Wheeler Cafarelli, “How Theories of Romanticism Exclude Women: Radcliffe, Milton, and the Legitimation of the Gothic Novel,” in *Milton, the Metaphysicals, and Romanticism*, ed. Lisa Low and Anthony John (Cambridge: Cambridge University Press, 1994), 84–113; Stuart Curran, “Charlotte Smith and British Romanticism,” *South Central Review* 11, no. 2 (1994): 66–78; Elizabeth Eger, *Bluestockings: Women of Reason from Enlightenment to Romanticism* (London: Palgrave Macmillan, 2010); Carol Shiner Wilson, *Re-Visioning Romanticism: British Women Writers, 1776–1837* (Philadelphia: University of Pennsylvania Press, 1994).

79. The phrase “the neo-Epicurean revival” comes from Richard W. F. Kroll, *The Material World: Literate Culture in the Restoration and Early Eighteenth Century* (Baltimore: Johns Hopkins University Press, 1991), 27. See also Mary Helen McMurran and Alison Conway, eds., *Mind, Body, Motion, Matter: Eighteenth-Century British and French Literary Perspectives* (Toronto: University of Toronto Press, 2016); Catherine Wilson, *Epicureanism at the Origins of Modernity* (Oxford: Oxford University Press, 2008).

80. Charles Taylor, *A Secular Age* (Cambridge: Cambridge University Press, 2007), 37–38.

81. Longinus, *On the Sublime*, 3.

82. Martin D. Yaffe, “‘Anthropogenic Effects’ in Genesis 1–11 and Francis Bacon,” *Telos* 177 (2016): 16, 19.

83. Haraway, *Modest_Witness@Second_Millenium*, 24.

84. As I show in “Mingling with Matter,” not all Enlightenment thinkers accepted the effectiveness of such transcendence.

85. Donna Haraway, “Anthropocene, Capitalocene, Plantationocene, Chthulucene, 160.

86. Haraway, *Modest_Witness@Second_Millennium*, 36.

87. Timothy Morton, *The Ecological Thought* (Cambridge, MA: Harvard University Press, 2010), 3.

Chapter One

1. Margaret Cavendish, *Poems, and Fancies* (London, 1653), 198.

2. Lucy Hutchinson, dedicatory epistle to *Lucretius de Rerum natura*, *The Works of Lucy Hutchinson*, vol 1, ed. Reid Barbour and David Norbrook (Oxford: Oxford University Press, 2012), 7.

3. David Norbrook, introduction to Hutchinson, *Lucretius de Rerum natura*, xxxiii.

4. Ibid., xxxiii–xlii.

5. Ibid., liv.

6. Ibid., xxxiii.

7. Steven Shapin and Simon Schaffer, *Leviathan and the Air-Pump: Hobbes, Boyle, and the Experimental Life* (Princeton, NJ: Princeton University Press, 1985),, 65–69.

8. Stacy Alaimo, *Bodily Natures: Science, Environment, and the Material Self* (Bloomington: Indiana University Press, 2010), 2.

9. Francis Bacon, *Novum Organum*, in *The Works of Francis Bacon*, ed. James

Spedding, Robert Leslie Ellis, and Douglas Denon Heath, vol. 8 (New York: Hurd and Houghton, 1864),, 76.

10. Ibid., 99

11. Shapin and Schaffer, *Leviathan*, 69

12. Ibid., 65–7; Robert Boyle, "Proëmical Essay . . . with some Considerations touching Experimental Essays in General," in *The Works of the Honourable Robert Boyle*, vol. 1, 2nd ed., ed. Thomas Birch (London: J. & F. Rivington, 1772), 301–7, 300.

13. Shapin and Schaffer, *Leviathan*, 69; Bacon, *Novum Organum*, 45.

14. Bacon, *Novum Organum*, 77.

15. Tita Chico, *The Experimental Imagination: Literary Knowledge and Science in the British Enlightenment* (Stanford, CA: Stanford University Press, 2018),; Lorraine Daston and Peter Galison, *Objectivity* (Brooklyn: Zone Books, 2007); Donna J. Haraway, *Modest_Witness@Second_Millenium.FemaleMan_Meets_OncoMouse: Feminism and Technoscience* (New York: Routledge, 1997).

16. In his *History of the Royal Society*, Thomas Sprat comments that the Royal Society "embrac[es] all assistance"—"not only by the Hands of learned and profess'd Philosophers: but from the Shops of *Mechanics*; from the Voyages of *Merchants*; from the Ploughs of *Husbandmen*; from the Sports, the Fishponds, the Parks, the Gardens of *Gentlemen*." Sprat, *History of the Royal Society* [1667], ed. Jackson I. Cope and Harold Whitmore (St. Louis: Washington University Press, 1958), 67, 72.

17. On the challenges that such principles posed to Baconian modesty for prominent philosophers including Boyle, Locke, and Hume, see esp. Helen Thompson, *Fictional Matter: Empiricism, Corpuscles, and the Novel* (Philadelphia: University of Pennsylvania Press, 2017), Jay L. Garfield, *The Concealed Influence of Custom: Hume's* Treatise *from the Inside Out* (Oxford: Oxford University Press, 2019).

18. Haraway, *Modest_Witness*, 27.

19. Ibid., 30.

20. Ibid, 31.

21. Hutchinson, dedicatory epistle to *Lucretius de Rerum natura*, 7.

22. Monte Ransome Johnson and Catherine Wilson, "Lucretius and the History of Science," in *The Cambridge Companion to Lucretius*, ed. Stuart Gillespie and Philip R. Hardie (Cambridge: Cambridge University Press, 2007), 132.

23. Hutchinson, *Lucretius*, lines 421–23, page 41.

24. Ibid., line 179, page 29.

25. Ibid., lines 60–64, page 87.

26. Ibid., line 93, page 89.

27. Ibid., dedicatory epistle to Lucretius, 7.

28. Boyle, "Proëmical Essay," 318.

29. I would suggest that this is one reason Bacon distanced himself from Epicureanism, despite his attraction to some of its foundational principles. Ransome and Wilson, "Lucretius and the History of Science," 135.

30. Lucy Hutchinson, preface to *Order and Disorder*, ed. David Norbrook (Oxford: Blackwell Publishing), 3.

31. This scholarship has challenged Reid Barbour's assertion that "[w]hatever

uses she may have derived from her work on Lucretius are largely cancelled after the Restoration." Reid Barbour, "Lucy Hutchinson, Atomism and the Atheist Dog," *Women, Science and Medicine 1500–1700: Mothers and Sisters of the Royal Society*, ed. Lynette Hunter and Sarah Hutton (Thrupp, Stroud, Gloucestershire: Sutton Pub., 1997), 123.

32. Alvin Snider, "Hutchinson and the Lucretian Body," *The New Science and Women's Literary Discourse: Prefiguring Frankenstein*, ed. Judy A. Hayden (New York, NY: Palgrave Macmillan, 2011), 31.

33. Snider, "Hutchinson," 31.

34. Mihoko Suzuki, "Animals and the Political in Lucy Hutchinson and Margaret Cavendish," *The Seventeenth Century* 30, no. 2 (April 3, 2015): 229.

35. Cassandra Gorman, "Lucy Hutchinson, Lucretius and Soteriological Materialism," *Seventeenth Century* 28, no. 3 (September 2013): 296.

36. Jonathan Goldberg, *The Seeds of Things: Theorizing Sexuality and Materiality in Renaissance Representations* (New York: Fordham University Press, 2009), 154; Reid Barbour, "Between Atoms and the Spirit: Lucy Hutchinson's Translation of Lucretius," Renaissance Papers (1994): 1–16; Barbour, "Lucy Hutchinson, Atomism and the Atheist Dog," 122–37.

37. Hutchinson, *Order and Disorder*, line 55, page 108.

38. Ibid., lines 356–57, page 99.

39. Ibid., line 110, page 110.

40. Ibid., lines 610–11, page 106.

41. Ibid., lines 441–44, page 119.

42. My emphasis; Hutchinson, *Lucretius*, line 173, page 27.

43. Ibid., lines 319–21, page 35.

44. Ibid., lines 286–95, page 35.

45. Hutchinson, *Order and Disorder*, lines 411–26, page 118–19.

46. See 1 Peter 3:20–21 (King James Version) for a biblical depiction of the deluge as a kind of baptism: "Which sometime were disobedient when once the longsuffering of God waited in the days of Noah, while the ark was a preparing, wherein few, that is, eight souls were saved by water. The like figure whereunto *even* baptism doth also now save us (not the putting away of the filth of the flesh, but the answer of a good conscience toward God,) by the resurrection of Jesus Christ."

47. Hutchinson, *Order and Disorder*, line 95, page 242.

48. Ibid., line 209, page 189.

49. Ibid., lines 83–85, page 242.

50. Hutchinson, *Lucretius*, line 173, page 27; lines 513–14, page 47; lines 541–52, page 49.

51. Hutchinson, *Order and Disorder*, lines 154, page 111.

52. Ibid., 111n 215.

53. Ibid., line 5, page 6.

54. Epicurus, "Letter to Menoeceus," in *The Epicurus Reader: Selected Writings and Testimonia*, ed. and trans. Brad Inwood and L. P. Gerson (Indianapolis: Hackett, 1994), 30.

55. Hutchinson, *Lucretius*, lines 1245–51, page 367.

56. Epicurus, "Letter to Menoeceus," 30.

57. Ibid., 31.

58. Lucy Hutchinson, *The Life of Mrs. Lucy Hutchinson*, in *Memoirs of the Life of Colonel Hutchinson with the fragment of an autobiography of Mrs. Hutchinson*, ed. James Sutherland (London: Oxford University Press, 1973), 286–87.

59. Hutchinson, *Life*, 264.

60. Norbrook, introduction to *Lucretius*, xxvii.

61. Hutchinson, preface to *Order and Disorder*, 4.

62. On the interrelatedness of science, religion, and literature in early modern England, see Kristin Girten, "Unsexed Souls: Natural Philosophy as Transformation in Eliza Haywood's *Female Spectator*," *Eighteenth-Century Studies* 43, no. 1 (2009): 55–74; Peter Harrison, "Sentiments of Devotion and Experimental Philosophy in Seventeenth-Century England," *Journal of Medieval and Early Modern Studies* 44, no. 1 (2014): 113–33; Courtney Weiss Smith, *Empiricist Devotions: Science, Religion, and Poetry in Early Eighteenth-Century England* (Charlottesville: University of Virginia Press, 2016).

63. William Derham, "To the Reader," *Physico-Theology: Or, a Demonstration of the Being and Attributes of God, from his Works of Creation*, 4th ed (London, 1716), n.p.

64. Robert Boyle, *Christian Virtuoso, Second Part, The Works of the Honourable Robert Boyle*, vol. 6 (London, 1772), 776.

65. Letter from Lord Coleraine to William Courten (April 1694), Sloane MS 3962, fos. 233–4, quoted in Katie Whitaker, "The Culture of Curiosity," *Cultures of Natural History*, ed. Nick Jardine, James A. Secord, and E. C. Spary (Cambridge: Cambridge University Press, 1996), 82.

66. "Physico-theology," *Oxford English Dictionary Online*.

67. See Harrison, "Sentiments of Devotion," 171; C. E. Raven, *Natural Religion and Christian Theology* (Cambridge: Cambridge University Press, 1953); Jeal Gillespie, "Natural History, Natural Theology, and Social Order: John Ray and the 'Newtonian Ideology,'" *Journal of the History of Biology* 20, no. 1 (1987): 1–49; Charles Webster, *The Great Instauration: Science, Medicine, and Reform, 1626–1660* (London: Duckworth, 1975), 150–51, 507–10.

68. Francis Bacon, *The Great Instauration*, in *Works*, vol. 8, 53.

69. Nehemiah Grew, "Epistle Dedicatory to the Right Reverend John Lord Bishop of Chester," *The Anatomy of Plants: With an Idea of a Philosophical History of Plants and Several Other Lectures Read Before the Royal Society* (London, 1682), n.p.

70. See for instance Suzuki, "Animals and the Political"; Goldberg, "Margaret Cavendish and Lucy Hutchinson: Writing Matter," 122–178; David Norbrook, "Margaret Cavendish and Lucy Hutchinson: Identity, Ideology and Politics," *In-Between: Essay & Studies in Literary Criticism* 9 (2000): 179–203; Emma L. E. Rees, "'A Horrible Precipice': Lucy Hutchinson's Lucretius," *Margaret Cavendish: Gender, Genre, Exile* (Manchester: Manchester University Press, 2003), 190–6.

71. Virginia Woolf, *The Essays of Virginia Woolf*, vol. 3: 1919–1924, ed. Andrew McNeillie (London: The Hogarth Press, 1988), 145.

72. Norbrook, "Margaret Cavendish and Lucy Hutchinson," 182.

73. On Cavendish's visit to the Royal Society, see Anna Battigelli, *Margaret Cavendish and the Exiles of the Mind* (Lexington: University Press of Kentucky, 1998), 110–13; Samuel Mintz, "The Duchess of Newcastle's Visit to the Royal Society," *Journal of English and Germanic Philology* 51 (1952): 168–76.

74. Londa Schiebinger, "Margaret Cavendish, Duchess of Newcastle," in *A History of Women Philosophers*, ed. Mary Ellen Waithe (Dordrecht: Kluwer Academic Publishers, 1991), 1–20; Stephen Clucas, "The Atomism of the Cavendish Circle: A Reappraisal," *The Seventeenth Century* 9, no. 2 (1994): 259–60. See also Liza Blake, "After Life in Margaret Cavendish's Vitalist Posthumanism," *Criticism* 62, no. 3 (2020): 433–56; Anne M. Thell, "'[A]s lightly as two thoughts': Motion, Materialism, and Cavendish's *Blazing World*," *Configurations* 23 (2015): 1–33. For an alternative view, see Lisa Sarasohn, "A Science Turned Upside Down: Feminism and the Natural Philosophy of Margaret Cavendish," *Huntington Library Quarterly* 47, no. 4 (1984): 299–307; John Rogers, *The Matter of Revolution: Science, Poetry and Politics in the Age of Reason* (Ithaca, NY: Cornell University Press, 1996), 188.

75. Stephen Clucas, "The Atomism of the Cavendish Circle," 259–60.

76. Ibid., 260.

77. Ibid., 261.

78. Ibid., 261.

79. My analysis builds on that of Emma Rees who argues that "the genre and structure of [Lucretius's] work facilitated [Cavendish's] own imaginative incursions into a print culture which, in being predominantly male and anti-Royalist, both reflected and was a product of the exiles under which Cavendish wrote in the 1650s." However, whereas Rees focuses on Cavendish's claim to literary authority, I focus instead on her claim to philosophical authority. Emma L. E. Rees, "'Sweet Honey of the Muses': Lucretian Resonance in *Poems, and Fancies*," *In-Between: Essays & Studies in Literary Criticism*, 9, no. 1 & 2 (2000): 3–16.

80. On Cavendish's critique of the Royal Society's practices, see especially Battigelli, *Margaret Cavendish*, 85–113; Lisa Anscomb, "'A Close, Naked, Natural Way of Speaking': Gendered Metaphor in the Texts of Margaret Cavendish and the Royal Society," *In-Between: Essays and Studies in Literary Criticism* 9, no. 1–2 (2000): 161–77; Kate de Rycker, "A World of One's Own: Margaret Cavendish and the Science of Self-Fashioning," *English Literature and the Disciplines of Knowledge, Early Modern to Eighteenth Century: A Trade for Light* (Boston: Rodopi, 2018), 76–93; Peter Dear, "A Philosophical Duchess: Understanding Margaret Cavendish and the Royal Society," *Science, Literature, and Rhetoric in Early Modern England* (Aldershot, England: Ashgate, 2007), 124–142; Eve Keller, "Producing Petty Gods: Margaret Cavendish's Critique of Experimental Science," *English Literary History* 64, no. 2 (1997): 447–71; Richard Nate, "'Plain and Vulgarly Express'd': Margaret Cavendish and the Discourse of the New Science," *Rhetorica: A Journal of the History of Rhetoric* 19, no. 4 (2001): 403–17; Denise Tillery, "'English Them in the Easiest Manner You Can': Margaret Cavendish on the Discourse and Practice of Natural Philosophy," *Rhetoric Review* 26, no. 3 (2007): 268–85; Jo Wallwork, "Disruptive Behaviour in the Making of Science: Cav-

endish and the Community of Seventeenth-Century Science," *Early Modern Englishwomen Testing Ideas*, ed. Jo Wallwork and Paul Salzman (Farnham, England: Ashgate, 2011), 41–53; Jo Wallwork, "Old Worlds and New: Margaret Cavendish's Response to Robert Hooke's *Micrographia*," *Women Writing, 1550–1750* (Bundoora, Australia: Meridian, 2001), 191–200.

81. According to Chico, the chemist Robert Boyle "takes Hooke's figuration a step further by removing the human body altogether." Chico, *Experimental Imagination*, 39.

82. Margaret Cavendish, *Observations upon Experimental Philosophy*, ed. Eileen O'Neill (Cambridge: Cambridge University Press, 2001), 50–51.

83. Ibid., 51.

84. Keller, "Producing Petty Gods," 451.

85. Ibid., 455.

86. Bacon, *Novum Organum*, 45.

87. Ibid., 44–45.

88. Ibid., 44.

89. Ibid., 44.

90. Ibid., 60–61.

91. Ibid., 77.

92. Ibid., 99.

93. Cavendish, *Observations*, 52.

94. Sprat, *History*, 3.

95. Cavendish, *Observations*, 67.

96. Blake, "After Life," 444. See also Keller, "Producing Petty Gods."

97. Cavendish, *Observations*, 56; Lucy Hutchinson, *Lucretius de Rerum natura*, line 179, page 29. Though there is no definitive evidence that Margaret Cavendish was familiar with Hutchinson's translation of Lucretius, which was begun in the early 1650s but not published until 1675, I have nevertheless chosen to use this translation because (1) it appears to have been the first English translation of *De Rerum Natura*, and (2) I believe there are connections to be made between Hutchinson's and Cavendish's interests in Lucretian Epicureanism and the gender implications of these interests. (David Norbrook even suggests that Hutchinson's translation may have been an attempt to compete with Cavendish.) Cottegnies provides a useful overview of the various Epicurean works that appear to have directly influenced Cavendish. Norbrook, introduction to *Lucretius de Rerum natura* xxxiii-xlii; Line Cottegnies, "Generic Bricolage and Epicureanism in Margaret Cavendish's Imaginative Works," *A Companion to the Cavendishes*, ed. Lisa Hopkins and Tom Rutter (York, United Kingdom: Arc Humanities Press, 2020), 327–28.

98. Margaret Cavendish, *The Grounds of Natural Philosophy* (London, 1668), 17. See also Cavendish, *Observations*, 35.

99. Hutchinson, *Lucretius*, lines 60–64, page 87.

100. Cavendish, *Grounds*, 30–31.

101. Hutchinson, *Lucretius*, line 93, page 89.

102. Liza Blake shows how Cavendish's belief in "the commonality of matter across creatures" allows her to, in the tradition of Lucretius, encourage her readers to "be happy about death" rather than to fear it. Blake, "After Life," 444; see also Hanlon, "Margaret Cavendish's Anthropocene Worlds," 64.

103. Cavendish, *Observations*, 56.

104. Ibid., 49.

105. Ibid., 95, 49.

106. Bacon, *The Great Instauration*, 48, 71.

107. Cavendish, *Observations*, 49.

108. Keller, "Producing Petty Gods," 457.

109. Cavendish, *Observations*, 53.

110. Sarasohn, *The Natural Philosophy of Margaret Cavendish*, 160; Cavendish, *Observations*, 141.

111. Cavendish, *Observations*, 61.

112. Though it is true, as Anna Battigelli observes, that Cavendish's recourse to reason distinguishes Cavendish's program from Baconian empiricism, I would challenge her polarization of their methods. Battigelli, *Margaret Cavendish*, 88–89.

113. Cavendish, *Observations*, 61.

114. Ibid., 61.

115. Ibid., 62.

116. Ibid., 62.

117. Ibid., 62.

118. Margaret Cavendish, *The Blazing World & Other Writings* (London: Penguin, 1994), 188–9. John Rogers presents a persuasive argument that Cavendish saw, in reason, the possibility of an escape from "the tyranny of sexual subjection." However, I would suggest that he neglects to honor the challenge Cavendish poses to binaristic thinking and, thus, extends his argument too far when he contends that she identifies women with rational matter and men with "the sensitive particles of matter." John Rogers, "Margaret Cavendish and the Gendering of the Vitalist Utopia," in *The Matter of Revolution: Science, Poetry, and Politics in the Age of Milton* (Ithaca, NY: Cornell University Press, 1996), 201–2. Keller too asserts that *The Blazing World* takes aim at such binaristic thinking. Keller, "Producing Petty Gods," 461.

119. Margaret Cavendish, *The Worlds Olio* (London, 1671), 42.

120. Ibid., 45.

121. Cavendish, *Observations*, 49.

122. "Trace," *Oxford English Dictionary Online*. On the centrality of tactility in Cavendish's writing and philosophy, see Misty G. Anderson, "Tactile Places: Materializing Desire in Margaret Cavendish and Jane Barker," *Textual Practice* 13, no. 2 (1999): 329–52.

123. Cavendish, *Worlds Olio*, 44.

124. "Sensitive," *Oxford English Dictionary Online*.

125. Cavendish, *Observations*, 62.

126. Cavendish, *Poems, and Fancies*, 32.

127. Ibid., 45.

128. On the association between philosophical practice and trivialities, see Kristin M. Girten, "Trivial Wonders: Fascination with the Negligible in British Literature of the Long Eighteenth Century," PhD diss., (Rutgers University, 2007).

129. Jonathan Lamb, "The Sublime," *The Cambridge History of Literary Criticism, IV: The Eighteenth Century*, ed. H. B. Nisbet and Claude Rawson (Cambridge: Cambridge University Press, 1997), 394–416.

130. Nicolas Boileau, *Traité du Sublime ou du merveilleux dans le discours* (Paris, 1674), ii.

131. This attribution has a long history. See especially Timothy Sean Quinn, "Anthropocene Crises and the Origin of Modernity" *Telos* 177 (winter 2016): 43–60; Sabine Wilke, "Enlightenment, Dialectic, and the Anthropocene: Bruised Nature and the Residues of Freedom" *Telos* 177 (winter 2016): 83–106; Jon Wittrock and Richard Polt, "Introduction: The Nature of Nature and the Politics of Fate" *Telos* 177 (winter 2016): 3–15; Martin D. Yaffe, "'Anthropogenic Effects' in Genesis 1–11 and Francis Bacon" *Telos* 177 (winter 2016): 16–42. Important works that helped establish the foundation for this attribution are Max Horkheimer and Theodor W. Adorno, *Dialectic of Enlightenment*, trans. John Cumming (London: Continuum, 2000), *Dialektik der Aufklarung* (New York: Social Studies Association, 1944); Carolyn Merchant, *The Death of Nature: Women, Ecology, and the Scientific Revolution* (New York: Harper & Row, 1980).

132. Bacon's relationship to Epicureanism is a source of debate among scholars. Robert Kargon argues that Bacon ultimately abandoned Epicureanism altogether. Reid Barbour characterizes Bacon's response to Epicurean materialism as "ambivalent." Graham Rees argues, "[t]here never was a period during which [Bacon] was, in any real sense, an atomic philosopher"; however, he nevertheless maintains that Epicureanism provided an important foundation for Bacon's method of attending to the "subtlety" of matter. Reid Barbour, *English Epicures and Stoics: Ancient Legacies in Early Stuart Culture* (Amherst: University of Massachusetts Press, 1998); Graham Rees, "Atomism and 'subtelty' in Francis Bacon's Philosophy," *Annals of Science* 37 (1980): 553, 558. See also Johnson and Wilson, "Lucretius and the History of Science"; Robert Kargon, "Walter Charleton, Robert Boyle, and the Acceptance of Epicurean Atomism in England," *Isis* 55, no. 2 (1964): 184–92; Graham Rees, "Matter Theory: A Unifying Factor in Bacon's Natural Philosophy?" *Ambix* 24, no. 2 (1977): 110–25.

133. Keller, "Producing Petty Gods," 450.

Chapter Two

1. For prints of the scenes Behn adapted, see Evariste Gheradi, *Théâtre italien, ou Recueil de toutes les scènes françaises qui ont été jouées sur le théâtre italien de l'Hôtel de Bourgogne* (Paris, 1741). On the literary and historical context of Behn's *Emperor*, see Janet Todd, *The Secret Life of Aphra Behn* (London: André Deutsch Ltd., 1996), 363–67.

2. Aphra Behn, *The Emperor of the Moon* (Oxford: Oxford University Press, 2008), 283.

3. Though Behn likely composed the play in 1684, it was not performed until 1687. On the play's composition and performance history, see Jane Spencer, introduction to *The Rover and Other Plays* (Oxford: Oxford University Press), xviii–xix. Richard W. F. Kroll, *The Material World: Literate Culture in the Restoration and Early Eighteenth Century* (Baltimore, MD: Johns Hopkins University Press, 1991), 27.

4. Aphra Behn, *Poems upon Several Occasions with, A Voyage to the Island of Love* (London, 1684), 50–57.

5. Alvin Snider, "Atoms and Seeds: Aphra Behn's Lucretius," *Clio* 33, no. 1 (Fall 2003): 6.

6. Ibid., 1–24.

7. Sarah-Gray Lesley, "The Epicure in Surinam: Lucretian Reception and Skepticism in Aphra Behn's Oroonoko," *Restoration: Studies in English Literary Culture, 1660–1700* 43, no. 2 (Fall 2019): 75–91; Mary Trull, "Lucretius's Performing Simulacra and Aphra Behn's 'Unresistable Idea' in *Love-Letters between a Nobleman and His Sister*," in *The Public's Open to Us All: Essays on Women and Performance in Eighteenth-Century England*, ed. Laura Engel and Linda V. Troost, 176–200 (Newcastle upon Tyne, England: Cambridge Scholars, 2009).

8. Ros Ballaster, "Taking Liberties: Revisiting Behn's Libertinism," *Women's Writing* 19, no. 2 (May 5, 2012): 165–76.

9. Robert Markley, "Global Analogies: Cosmology, Geosymmetry, and Skepticism in Some Works of Aphra Behn," in *Science, Literature and Rhetoric in Early Modern England*, ed. Juliet Cummins and David Burchell (Hampshire, England: Ashgate, 2007), 206.

10. Ibid., 206.

11. Ibid., 212.

12. Aphra Behn, "To the Unknown Daphnis on his Excellent Translation of *Lucretius*," in *The Works of Aphra Behn*, ed. Janet Todd (London: William Pickering, 1992), line 6, page 25.

13. Ibid., lines 45–51, page 26.

14. Thomas Creech preface to *T. Lucretius Carus, the Epicurean Philosopher, His Six Books* De Natura Rerum *Done into English Verse*, trans. by Thomas Creech (Oxford, 1682), n.p.

15. Trull, "Lucretius's Performing Simulacra," 193.

16. Ibid., 177.

17. Ibid., 178.

18. Ibid., 179.

19. Ibid., 179.

20. Ibid., 180–2; Lucretius, *De Natura Rerum*, trans. Creech, 133–34.

21. Ballaster, "Taking Liberties," 166.

22. Al Coppola, "Retraining the Virtuoso's Gaze: Behn's Emperor of the Moon, The Royal Society, and the Spectacles of Science and Politics," *Eighteenth-Century Studies* 41, no. 4 (2008 Summer 2008): 482, 498.

23. Ibid., 492, 483.

24. Ibid., 486.

25. Behn, *Emperor*, 283.

26. Nehemiah Grew, *Musaeum Regalis Societatis or a Catalog and Descriptions of the Natural and Artificial Rarities Belonging to the Royal Society and Preserved at Gresham College* (London, 1681), 360–4.

27. Ibid., dedicatory epistle, n.p.

28. Behn, *Emperor*, 283.

29. Grew, *Musaeum*, 1, 17, 26, 158, 215.

30. Coppola, "Retraining the Virtuoso's Gaze," 487–88.

31. Grew, *Musaeum*, 9–10.

32. Behn, *Emperor*, 284.

33. Ibid., 285.

34. Ibid., 285.

35. Ibid., 285.

36. Later in this chapter, we will explore how such language of "ravishment" associates matter's seductiveness and capacity to discompose with the sublime.

37. Behn, *Emperor*, 286.

38. My use of the term *charm* here is informed by Ballaster, "Taking Liberties," 165–76. I address the concept fully below.

39. Behn, *Emperor*, 291–96.

40. Ibid., 295. For a relevant study of the history of the closet, see Danielle Bobker, *The Closet: The Eighteenth-Century Architecture of Intimacy* (Princeton, NJ: Princeton University Press, 2020).

41. Behn, *Emperor*, 290.

42. Ibid., 290.

43. Ibid., 290–91.

44. Ibid., 299.

45. Ibid., 289.

46. "Ravished," *Oxford English Dictionary Online*.

47. Behn, *Emperor*, 287.

48. Ibid., 288.

49. "Tickle," *Oxford English Dictionary Online*.

50. Behn, *Emperor*, 305.

51. Lucretius, *De Natura Rerum*, trans. Creech, 132.

52. Trull, "Lucretius's Performing Simulacra," 181.

53. Behn, *Emperor*, 332.

54. Ibid., 332.

55. Ibid., 333.

56. Mopsophil expresses libertinism earlier in the play when, in the first act, she comments: "Well, 'tis a delicious thing to be rich; what a world of lovers it invites: I have one for every hand, and the favourite for my lips." Behn, *Emperor*, 292.

57. Behn, "Daphnis," *Works*, lines 2–3, page 25.

58. Ibid., lines 4–16, page 25.

59. Ibid., lines 17–20, page 25.

60. Snider, "Atoms and Seeds," 3.

61. Behn, "Daphnis," *Works*, lines 41–44, page 26.

62. Ibid., lines 25–28, page 25.

63. Ibid., lines 29–34, pages 25–26.

64. Ibid., lines 45–56, page 26.

65. Aphra Behn, preface to Bernard le Bovier de Fontenelle, *A Discovery of New Worlds*, trans. Aphra Behn (London, 1688), n.p.

66. Ibid., translator's preface, n.p.

67. Ibid., dedicatory epistle, n.p.

68. Ibid., n.p.

69. According to Vivian Appler, "A focus on her late writing suggests an engagement with science concepts, objects, and philosophers that renders Behn at the very least a competent, critical, credible participant in the late seventeenth-century 'new science' conversation, if not a natural philosopher in her own right." Therefore, Behn "should be recognized a dynamic female contributor to the history of science." Vivian Appler, "'Shuffled Together under the Name of a Farce': Finding Nature in Aphra Behn's *The Emperor of the Moon*," *Theatre History Studies* 37 (2018): 28. On Behn's philosophical authority, see also Markley, "Global Analogies"; Sarah Goodfellow, "'Such Masculine Strokes': Aphra Behn as Translator of 'A Discovery of New Worlds,'" *Albion: A Quarterly Journal concerned with British Studies* 28, no. 2 (Summer 1996): 229–50.

70. Aphra Behn, dedicatory epistle to Fontenelle, *A Discovery of New Worlds*, trans Behn, n.p.

71. Fontenelle, author's preface to *Discovery*, n.p.

72. Fontenelle, *Discovery*, 25.

73. Ibid., 3.

74. Ibid., 3–4.

75. Fontenelle, To Monsieur de L——, *Discovery*, n.p.

76. Fontenelle, author's preface to *Discovery*, n.p.

77. Fontenelle, *Discovery*, 39.

78. Ibid., 39.

79. Ibid., 39–40.

80. Ibid., 71–72.

81. Sarah-Gray Lesley has recently published a valuable study of the influence of Lucretian Epicureanism on *Oroonoko*. Sarah-Gray Lesley, "The Epicure in Surinam: Lucretian Reception and Skepticism in Aphra Behn's Oroonoko," *Restoration: Studies in English Literary Culture, 1660–1700* 43, no. 2 (Fall 2019): 75–91.

82. Aphra Behn, *Oroonoko*, ed. Joanna Lipking (New York: Norton, 1997), 8.

83. Ibid., 8.

84. Ballaster, "Taking Liberties," 166.

85. Behn, *Oroonoko*, 14–15.

86. My emphasis; ibid., 16.

87. Ibid., 12.

88. Ibid., 56.

89. Ibid., 51.

90. According to Gallagher, "There is . . . a close connection between Behn's sustained authorial presence in this book . . . and the black hero's lustre; as the story moves forward, a mutual polishing takes place in which narrator and hero buff off each other's obscurity." Catherine Gallagher, "Oroonoko's Blackness," in *Aphra Behn Studies*, ed. Janet Todd, (Cambridge: Cambridge University Press, 1996), 237–238.

91. Ibid., 239.

92. Ramesh Mallipeddi, "Spectacle, Spectatorship, Sympathy: Aphra Behn's *Oroonoko* and the English Commercial Empire," in *Spectacular Suffering: Witnessing Slavery in the Eighteenth-Century British Atlantic* (Charlottesville: University of Virginia Press, 2016), 25–50.

93. Ibid., 41.

94. Laura Brown, "The Romance of Empire: *Oroonoko* and the Trade in Slaves," *The New Eighteenth Century: Theory, Politics, English Literature*, ed. Felicity Nussbaum and Laura Brown (London: Methuen Publishing, 1987), 47.

95. Ibid., 47.

96. Ibid., 48.

97. Lucretius, *On the Nature of the Universe*, trans. Ronald Melville (Oxford: Oxford University Press, 2009), 42. Thomas Creech portrays these same lines as follows, using the slightly less evocative "decline" to translate Lucretius' "clinamen":

> Now *Seeds* in downward motion must *decline*,
> Tho *very little* from th' exactest line;
> For did they still move *strait*, they needs must fall
> Like drops of Rain, dissolv'd and scatter'd all,
> For ever tumbling thro the mighty space,
> And never *joys* to make one single mass.

Lucretius, *De Natura Rerum*, trans. Creech, 41.

98. Rebekah Mitsein, "Trans-Saharan Worlds and World Views in Aphra Behn's *Oroonoko*," *Eighteenth-Century Fiction* 30, no. 3 (Spring 2018): 344–45.

99. Robert Boyle, *General Heads for the Natural History of a Country Great or Small Drawn out for the Use of Travellers and Navigators* (London, 1692), 1–2.

100. Ibid., 9.

101. Ibid., 7.

102. David N. Livingstone, *Putting Science in Its Place: Geographies of Scientific Knowledge* (Chicago: University of Chicago Press, 2003), 148.

103. Richard Jobson, *The Golden Trade: Or, A Discouery of the Riuer Gambra, and the Golden Trade of the Aethiopians* (London: 1623), 37–38.

104. Anon., *The Golden Coast, or, A Description of Guinney* (London: 1665), 76–77.

105. Mitsein demonstrates that it was common for such travel narratives to rely significantly on oral accounts provided by indigenous inhabitants for such details, a fact obscured by the tendency of modest witnesses to claim that their observations were firsthand. Mitsein, "Trans-Saharan Worlds," 339–68.

106. Tita Chico, *The Experimental Imagination: Literary Knowledge and Science in the British Enlightenment* (Stanford, CA: Stanford University Press, 2018), 37.

107. Donna J. Haraway, *Modest_Witness@Second_Millenium.FemaleMan_Meets_OncoMouse: Feminism and Technoscience* (New York: Routledge, 1997), 25.

108. My emphasis; Behn, *Oroonoko*, 14.

109. Ibid., 38.

110. Ibid., 63–64.

111. Ibid., 8.

112. Ibid., 41.

113. Ibid., 41.

114. Ibid., 41.

115. Suvir Kaul, "Reading Literary Symptoms: Colonial Pathologies and the *Oroonoko* Fictions of Behn, Southerne, and Hawkesworth," *Eighteenth-Century Life* 18, no. 3 (November 11, 1994): 82.

116. Ibid., 83.

117. Laura Doyle, "The Folk, the Nobles, and the Novel: The Racial Subtext of Sentimentality," *Narrative* 3, no. 2 (May 5, 1995): 172.

118. Behn, *Oroonoko*, 41.

119. Ibid., 41.

120. Doyle, "The Folk," 173.

121. Kaul, "Reading Literary Symptoms," 83.

122. Doyle, "The Folk," 173.

123. The introduction to this book discusses this notion of "property in one's person," as it is theorized by John Locke. John Locke, *Second Treatise of Government*, ed. C. B. Macpherson (Indianapolis, IN: Hackett Publishing Co., 1980), 19.

124. Jonathan Lamb, "The Sublime," in *The Cambridge History of Literary Criticism, IV: The Eighteenth Century*, ed. H. B. Nisbet and Claude Rawson (Cambridge: Cambridge University Press, 1997), 394.

125. Behn, *Oroonoko*, 65.

126. Ibid., 61.

127. Leonard Welsted, "Epistles, odes, &c. written on several subjects. With a translation of Longinus's treatise on the sublime" (London, 1724), 143.

128. Behn, *Oroonoko*, 63.

129. William Smith, *Dionysius Longinus on the Sublime* (Dublin, 1740), 3.

130. Behn, *Oroonoko*, 64.

131. Characterizing obscurity as a distinguishing feature of the sublime, Burke asserts, "No person seems better to have understood the secret of heightening, or of setting terrible things, if I may use the expression, in their strongest light by the force of a judicious obscurity, than Milton. His description of Death in the second book is admirably studied; it is astonishing with what a gloomy pomp, with what a significant and expressive uncertainty of strokes and colouring he has finished the portrait of the king of terrors." Edmund Burke, *A Philosophical Enquiry into the Origin of Our Ideas of the Sublime and Beautiful*, ed. James T. Boulton (London: Routledge and Paul, 1958), 59.

132. Behn, *Oroonoko*, 63.

133. John Dennis, *The Grounds of Criticism in Poetry* (London, 1704), 79.

134. This latter definition is now archaic. *Oxford English Dictionary Online.*

135. Behn, *Oroonoko*, 64.

136. Ibid., 60.

137. Ibid., 44.

138. Ibid., 46.

139. Markley, "Global Analogies," 212.

140. Kaul, "Reading Literary Symptoms," 88.

141. Behn refers to Oroonoko as "her slave" in the last sentence of the work's dedicatory epistle to Lord Maitland: "'Tis purely the Merit of my Slave that must render it worthy of the Honour it begs; and the Author of that of Subscribing herself." Behn, *Oroonoko*, 7.

Chapter Three

1. For a history of the Enlightenment "skeptical sublime," see James Noggle, *The Skeptical Sublime: Aesthetic Ideology in Pope and the Tory Satirists* (Oxford: Oxford University Press, 2001). See also Sarah Tindal Kareem, *Eighteenth-Century Fiction and the Reinvention of Wonder* (Oxford: Oxford University Press, 2014).

2. Mary Beth Harris, "Upsetting the Balance: Exposing the Myth of Masculine Virtue and Desire in Eliza Haywood's *Philidore and Placentia*," *The Eighteenth Century* 58, no. 1 (2017): 197.

3. Ibid., 198.

4. Ibid., 206.

5. Ibid., 206.

6. On desire in Haywood, see especially Harris, "Upsetting the Balance," 195–218; Mary Beth Harris, "Reforming Count D'elmont: Masculinity, Sympathy, and Reading in Eliza Haywood's *Love in Excess*," *Tulsa Studies in Women's Literature* 38, no. 2 (Fall 2019): 285–311; Paul Kelleher, *Making Love : Sentiment and Sexuality in Eighteenth-Century British Literature* (Lanham, Maryland: Bucknell University Press, 2015); Jonathan Brody Kramnick, "Locke, Haywood, and Consent," *English Literary History* 72 (2005): 453–70; Kathleen Lubey, "'Too Great Warmth": Joseph Addison, Eliza Haywood, and the Pleasures of Reading," *Excitable Imaginations: Eroticism and Reading in Britain, 1660–1760* (Lewisburg: Bucknell University Press, 2012), 71–108.

7. Eve Tavor Bannet, "Haywood's Spectator and the Female World," *Fair Philosopher: Eliza Haywood and the Female Spectator*, ed. Lynn Marie Wright and Donald J. Newman (Lewisburg, PA: Bucknell University Press, 2006), 83.

8. G. Gabrielle Starr, "Rereading Prose Fiction: Lyric Convention in Aphra Behn and Eliza Haywood." *Eighteenth-Century Fiction* 12, no. 1 (1999): 2.

9. Ibid., 2.

10. Eliza Haywood, *The Female Spectator, Volumes 1 and 2, Selected Works of Eliza Haywood*, ed. Alexander Pettit, vol. 2 (London: Pickering and Chatto, 2001), 17.

11. Joseph Addison and Sir Richard Steele, *The Spectator*, ed. Donald Bond, vol. 1 (Oxford: Clarendon Press, 1965), 4–5.

12. Ibid., 5, 19.

13. Ibid., 20.

14. Ronald Paulson, *The Beautiful, Novel, and Strange: Aesthetics and Heterodoxy* (Baltimore, MD: Johns Hopkins University Press, 1996), 50. As this quote from Paulson suggests, by 1711 when the first *Spectator* paper was published, precisely how gentlemen chose to practice modest witnessing varied according to their precise philosophical and/or political inclinations.

15. David Marshall, *The Frame of Art: Fictions of Aesthetic Experience, 1750–1815* (Baltimore, MD: Johns Hopkins University Press, 2005), 177.

16. Haywood, *Female Spectator*, vol. 2, 18.

17. As Paulson demonstrates, "[w]hile addressing himself ostentatiously to the issue of including the feminine, 'the female world,' . . . within his audience," his modest "refusal to involve himself . . . join[s] spectatorship to voyeurism." Moreover, according to Paulson, Addison and Steele "retain a polite version of the fraternal aspect of Shaftesbury's civic humanism" as "they transmit the Shaftesburian aesthetic of contempt for the 'interested' reproductivity of heterosexuality." Paulson, *Beautiful, Novel, and Strange*, 51–52.

18. "Engross," *Oxford English Dictionary Online*. As Marshall observes, it is through his attention to everyday life, and individuals' responses to it, that Addison develops his understanding of aesthetic experience. Marshall, *The Frame of Art*, 5–7. See also Jerome Stolnitz, "On the Origins of Aesthetic Disinterestedness," *Journal of Aesthetics and Art Criticism* 20 (1961): 137.

19. Addison and Steele, *Spectator*, vol. 1, 6.

20. Ibid., 7–13.

21. Ibid., 6.

22. Haywood, *Female Spectator*, vol. 2, 18.

23. Ibid., 18.

24. Ibid., 19.

25. Ibid., 19.

26. On *The Spectator* and the public sphere, see especially Terence Bowers, "Universalizing Sociability: The Spectator, Civic Enfranchisement, and the Rule(s) of the Public Sphere," in *The Spectator: Emerging Discourses*, ed. Donald J. Newman (Newark, DE: University of Delaware Press, 2005), 150–74; Brian Cowan, "Mr. Spectator and the Coffeehouse Public Sphere," *Eighteenth-Century Studies* 37, no. 3 (Spring 2004): 345–66; Jürgen Habermas, *The Structural Transformation of the Public Sphere: An Inquiry into a Category of Bourgeois Society* (Cambridge, MA: MIT Press, 1989); Markman Ellis, "Coffee-Women, The Spectator and the Public Sphere in the Early Eighteenth Century," in *Women, Writing and the Public Sphere, 1700–1830*, ed. Elizabeth Eger, Charlotte Grant, Clíona Ó Gallchoir, and Penny Warburton (Cambridge: Cambridge University Press, 2001), 27–52.

27. "Six and more cry with one voice." Addison and Steele, *Spectator*, vol. 1, 7.

28. Haywood, *Female Spectator*, vol. 2, 19.

29. On the frontispieces to *The Female Spectator*, see Janine Barchas, "Apollo, Sap-

pho, and—a Grasshopper?! A Note on the Frontispieces to *The Female Spectator*," in *Fair Philosopher*, 60–71.

30. Addison and Steele, *Spectator*, vol. 1, 6.

31. Ibid., 6.

32. Ibid., 5.

33. Haywood, *Female Spectator*, vol. 2, 19.

34. Ibid., 19.

35. Ibid., 19.

36. Ibid., 20.

37. Ibid., 20.

38. Ibid., 20.

39. Ibid., 19.

40. Ibid., 19.

41. Ibid., 19.

42. Paulson, *Beautiful, Novel, and Strange*, 51–52.

43. Haywood, *Female Spectator*, vol. 2, 127–35.

44. Ibid., 133.

45. Ibid., 133.

46. Ibid., 19.

47. Ibid., 134.

48. Jennie Batchelor and Manushag N. Powell, *Women's Periodicals and Print Culture in Britain, 1690–1820s: The Long Eighteenth Century* (Edinburgh: Edinburgh University Press, 2018), 23.

49. Haywood, *Female Spectator*, vol. 2, 18.

50. Adela Pinch, *Strange Fits of Passion: Epistemologies of Emotion, Hume to Austen* (Stanford, CA: Stanford University Press, 1996), 35.

51. David Hume, *A Treatise of Human Nature*, ed. L. A. Selby-Bigge; 2nd ed., ed. P. H. Nidditch (Oxford: Oxford University Press, 1978), 354.

52. On sympathy in Hume, see esp. Jay L. Garfield, *The Concealed Influence of Custom: Hume's* Treatise *from the Inside Out* (Oxford: Oxford University Press, 2019); Davide Panagia, *Impressions of Hume: Cinematic Thinking and the Politics of Discontinuity* (Lanham, MD: Rowman & Littlefield, 2013).

53. Haywood, *Female Spectator*, vol. 2, 19.

54. Ibid., 18.

55. Stephen Ahern, *Affected Sensibilities: Romantic Excess and the Genealogy of the Novel, 1680–1810* (New York: AMS Press, 2007), 87.

56. Jonathan Lamb, *The Evolution of Sympathy in the Long Eighteenth Century* (London; Brookfield, VT: Pickering & Chatto, 2009), 17.

57. Adam Smith, *The Theory of Moral Sentiments*, ed. D. D. Raphael and A. L. Macfie (Indianapolis: Liberty Fund, 1984), 9.

58. Ibid., 21.

59. Ibid., 25–26.

60. Though this difference may seem to suggest that Smith is surprisingly, on this particular issue, more skeptical than Hume, Lamb shows that in fact the opposite

is the case. As Lamb explains, "With Burke, Hume, and Mandeville all muddying the waters on the other side of the stream occupied by the proponents of the existential purity of the thinking thing and the infallible judgements of the moral sense, the questions concerning self, will and power needed some sort of adjudication and stabilization, and this was provided by Adam Smith." With his theory of the impartial spectator, Smith demonstrates that he has faith that the self can be thus adjudicated and stabilized. Hume has no such faith (nor does he believe it to be necessary for sympathy). Lamb, *Evolution*, 37. On sympathy in Hume and Smith, see also Elias L. Khalil, "The Fellow-Feeling Paradox: Hume, Smith and the Moral Order," *Philosophy* 90, no. 354 (2015): 653–78; Helga Schwalm, "Transformations and Migrations of Sympathy in Eighteenth-Century Moral Philosophy and Poetics," *Poetica* 47, no. ¾ (2015): 151–75. Silvia Sebastiani, "Beyond Ancient Virtues: Civil Society and Passions in the Scottish Enlightenment," *History of Political Thought* 32, no. 5 (2011): 821–40.

61. Fred Parker, "Addison's Modesty: Or, The Essayist as Spectator," in *Joseph Addison: Tercentenary Essays*, ed. Paul Davis (Oxford: Oxford University Press, 2021) 174.

62. Addison and Steele, *Spectator*, vol. 3, 243.

63. Ibid., 244.

64. Of the literary style of *The Spectator*, Parker writes, "The special quality of Addison's style is largely indicated [by contemporaries such as Hugh Blair and Samuel Johnson] through negatives, as though its genius lay precisely in its capacity to recede from view." Parker, "Addison's Modesty," 180.

65. Addison and Steele, *Spectator*, vol. 1, 6.

66. Haywood, *Female Spectator*, vol. 2, 18.

67. Addison prints the quote in its original Latin as he does with all such epigraphs: "Non fumum ex fulgore, sed ex fumo dare lucem / Cogitat, ut speciosa dehinc miracula promat." Addison and Steele, *Spectator*, vol. 1, 1.

68. Ibid., 1.

69. Ibid., 1.

70. Ibid., 5.

71. Ibid., 44.

72. Addison and Steele, *Spectator*, vol. 3, 539.

73. James Hodges was the first scholar to associate Haywood's *The Female Spectator* with the conduct-book tradition. James Hodges, "*The Female Spectator*, a Courtesy Periodical," *Studies in the Early English Periodical*, ed. Richmond P. Bond (Chapel Hill: University of North Carolina Press, 1957), 151–82. See also Patricia Meyer Spacks, Introduction to *Selections from* The Female Spectator *by Eliza Haywood*, ed. Patricia Meyer Spacks (Oxford: Oxford University Press, 1999), ix–xii.

74. Addison and Steele, *Spectator*, vol. 1, 44.

75. Ibid., 44.

76. "Philosophy," *Oxford English Dictionary*.

77. Addison and Steele, *Spectator*, vol. 1, 4–5.

78. Ibid., 5.

79. Lubey, "'Too Great Warmth,'" 89.

80. Addison and Steele, *Spectator*, vol. 3, 561.

81. In no. 418 of the Pleasures of the Imagination papers, Addison writes, "There is yet another Circumstance which recommends a Description more than all the rest, and that is, if it represents to us such Objects as are apt to raise *a secret Ferment* in the Mind of the Reader, and to work, with Violence, upon his Passions. For, in this Case, we are at once *warmed* and enlightned [*sic*], so that the Pleasure becomes more Universal, and is several ways qualified to entertain us." My emphasis; Addison and Steele, *Spectator*, vol. 3, 567.

82. Lubey does acknowledge that Addison seeks to "tame . . . the body from becoming *too* warm." She argues, "Since subjects' material substance will be warmed, perhaps inflamed by texts, they must be trained to aestheticize—to keep active, surprising, and mobile—those passions that are raised by beauty, passions that threaten to remain material, visceral, libidinal." Lubey, "'Too Great Warmth,'" 88–89.

83. On "virtual witnessing" and the Royal Society, see esp. Steven Shapin and Simon Schaffer, *Leviathan and the Air-Pump: Hobbes, Boyle, and the Experimental Life* (Princeton, NJ: Princeton University Press, 1985), 60–65.

84. Lubey, "'Too Great Warmth,'" 106.

85. Lubey, "'Too Great Warmth,'" 106.

86. Haywood, *Female Spectator*, vol. 2, 22.

87. Ibid., 22–23.

88. Ibid., 23–24.

89. Ibid., 24.

90. Ibid., 24.

91. Ibid., 27.

92. Ibid., 27.

93. Ibid., 28.

94. Ibid., 28.

95. Ibid., 20.

96. Ibid., 20.

97. Ibid., 18.

98. For a nuanced study of the techniques Haywood uses to help her readers control and moderate emotional excess, see Lubey, "'Too Great Warmth,'" 71–108.

99. Haywood, *Female Spectator*, vol. 2, 175.

100. Ibid., 175.

101. Ibid., 175–76.

102. Ibid., 176.

103. Ibid., 176.

104. Ibid., 179, 181–82.

105. Ibid., 176.

106. Ibid., 183.

107. Ibid., 182.

108. Ibid., 182.

109. As we observed above, according to the Spectator, philosophy prevents the "sprouting" of "Follies." Addison and Steele, *Spectator*, vol. 1, 44.

110. "Philosophy," *Oxford English Dictionary Online.*

111. "Empirical philosophy," *Oxford English Dictionary Online.*

112. Haywood, *Female Spectator,* vol. 2, 19.

113. This section of the chapter draws from my article Kristin Girten, "Unsexed Souls: Natural Philosophy as Transformation in Eliza Haywood's *Female Spectator,*" *Eighteenth-Century Studies* 43, no. 1 (2009): 55–74.

114. Haywood, *Female Spectator,* vol. 3, 82.

115. Ibid., 90.

116. Ibid., 83.

117. Ibid., 84.

118. Philo-Naturæ's use of fashion metaphors to portray other small insects has a similar effect. See Haywood, *Female Spectator,* vol. 3, 85. Scholarship in the history of women in science has made similar claims. See, for example, Patricia Phillips, *The Scientific Lady: A Social History of Women's Scientific Interests, 1520–1918* (London: Weidenfeld and Nicolson, 1990); Lynette Hunter and Sarah Hutton, eds., *Women, Science and Medicine, 1500–1700: Mothers and Sisters of the Royal Society* (Stroud, UK: Sutton Publishing, 1997); Ann B. Shteir, *Cultivating Women, Cultivating Science: Flora's Daughters and Botany in England, 1760–1860* (Baltimore, MD: Johns Hopkins University Press, 1996).

119. Haywood, *Female Spectator,* vol. 3, 85. As Phillips observes, silkworm propagation was historically considered a female activity. Phillips, *Scientific Lady,* 43–50.

120. Haywood, *Female Spectator,* vol. 3, 88–89.

121. Margo Collins suggests that many of Haywood's works invite such transvestism. Margo Collins, "Eliza Haywood's Cross-Gendered Amatory Audience," *Eighteenth-Century Women* 2, no. 1 (2002): 43–60.

122. Haywood, *Female Spectator,* vol. 3, 83.

123. Ibid., 161.

124. Ibid., 161. Phillips, Iliffe, and Willmoth have suggested that in seventeenth- and eighteenth-century England, the domestic retreat (particularly when situated in a rural setting) was at times perceived to possess natural affinities with philosophic practice. Phillips, *Scientific Lady*; Rob Iliffe and Frances Willmoth, "Astronomy and the Domestic Sphere: Margaret Flamsteed and Caroline Herschel as Assistant-Astronomers," in *Women, Science and Medicine,* 257–62.

125. Haywood, *Female Spectator,* vol. 3, 164.

126. Ibid., 164.

127. Ibid., 165.

128. Haywood's exploration of the significance of magnification to gender construction echoes two of Margaret Cavendish's poems, "Of many Worlds in this World" and "A World in an Eare-Ring" (1653). The latter poem elaborates the former's conclusion, which consists of the following couplet: "And if thus small, then ladies may well wear / A world of worlds, as pendents in each ear." Girten, "Unsexed Souls," 62. Margaret Cavendish, *Poems, and Fancies* (London, 1653), 44–46.

129. This effect becomes even more apparent when, in her response to Philo-

Naturæ's second letter, Haywood demonstrates the capacity of natural philosophy to make wonders out of feminine "Ornaments." Transformed by philosophy from baubles pleasing to the eye and nose into engines of religious instruction and devotion, the flowers women use to decorate their "Bosoms," "Hair," "Pots," and "*China* Basons" become truly valuable and therefore worthy of serious consideration. One must assume that so, too, do the women who wear them for, as Haywood acknowledges, women are so closely identified with the trivialities they employ to decorate themselves and their environs that it is impossible to separate them from one another. Haywood, *Female Spectator*, vol. 3, 245–50.

130. Ibid., 169.

131. Francis Bacon, *The New Organon*, trans. and ed. Lisa Jardine and Michael Silverthorne (New York: Cambridge University Press, 2000), 119.

132. Ibid., 92–93.

133. Ibid., 228–29.

134. Ibid., 228–29. Though, as Lorraine Daston argues, "strange facts" had their place in the new philosophy, the mundane also occupied an important—and I would argue even more prominent—position; see Lorraine Daston, "Baconian Facts, Academic Civility, and the Prehistory of Objectivity," *Annals of Scholarship* 8, nos. 3–4 (1991): 337–63.

135. Abraham Cowley, "Preface," in *The History of the Royal-Society of London for the Improving of Natural Knowledge*, by Thomas Sprat (London, 1667),, n.p.

136. On the importance of the mundane, mean, and diminutive within England's new philosophy, see Kristin M. Girten, "Trivial Wonders: Fascination with the Negligible in British Literature of the Long Eighteenth Century," PhD diss., (Rutgers University, 2007). For how the formal and historical connections between the new philosophy and the Christian tradition of meditation also lend authority to Haywood's gender egalitarianism, see Girten, "Unsexed Souls," 64–66.

137. Haywood, *Female Spectator*, vol. 3, 162.

138. Shawn Lisa Maurer maintains that Haywood's periodical "push[es] at the boundaries" of the female domain but does so without "reconfigur[ing] them"—that it "confine[s]" women "to a subset of the universal sphere designated male." In contrast, I would argue that to push at the boundaries of the female domain to the extent that Haywood does in *The Female Spectator* is necessarily to reconfigure them, as well as those of the male sphere. Shawn Lisa Maurer, "A Women's Magazine? Masculine Subjectivity in Periodicals for and by Women," in Shawn Lisa Maurer, *Proposing Men: Dialectics of Gender and Class in the Eighteenth-Century British Periodical* (Stanford, CA: Stanford University Press, 1998), 204–31.

139. Haywood's empirical activities thus offer an expression of what Mary Anne Schofield has portrayed as *The Female Spectator*'s "doctrine of seeming compliance but actual revolt." Mary Anne Schofield, *Eliza Haywood* (Boston: Twayne Publishers, 1985), 110.

140. See also Haywood's engagement with the "Plurality of Worlds" hypothesis, in which she similarly asserts her capacity to philosophize on an equal footing with men. Girten, "Unsexed Souls," 67–68.

141. Haywood, *Female Spectator*, vol. 3, 174.

142. Ibid., 174.

143. Ibid., 174–75.

144. Ibid., 174–75.

145. Ibid., 174.

146. Ibid., 174.

147. My emphasis; ibid., 174.

148. Helen Thompson, "Plotting Materialism: W. Charleton's *The Ephesian Matron*, E. Haywood's *Fantomina*, and Feminine Consistency," *Eighteenth-Century Studies* 35, no. 2 (Winter 2002): 195–214.

149. Haywood, *Female Spectator*, vol. 3, 173.

150. Ibid., 173.

151. Ibid., 174.

152. Ibid., 174.

153. Ibid., 174.

154. The term "engrossed" comes from a passage from *The Female Spectator* quoted above at note 16. Haywood, *Female Spectator*, vol. 2, 18.

155. Noggle, *The Skeptical Sublime*, 5.

156. For a foundational attack on "enthusiasm," see John Locke, "Of Enthusiasm," *An Essay Concerning Human Understanding*, ed. Roger Woolhouse (London: Penguin Books),, 614–22.

157. On the history of curiosity in the Enlightenment era, see Barbara M. Benedict, *Curiosity: A Cultural History of Early Modern Inquiry* (Chicago: University of Chicago Press, 2001).

158. Eliza Haywood, *Fantomina*, *Fantomina and Other Works*, ed. Alexander Pettit, Margaret Case Croskery, and Anna C. Patchias (Peterborough, Ontario: Broadview Press, 2004), 41.

159. My emphasis; Haywood, *Fantomina*, 41.

160. Ibid., 51.

161. Ibid., 51.

162. Haywood, *Female Spectator*, vol. 2, 17.

163. Ibid., 17–18.

164. Haywood, *Fantomina*, 71.

165. My emphasis; Haywood, *Female Spectator*, vol. 2, 18.

166. My emphasis; ibid., 18.

167. Haywood, *Female Spectator*, vol. 3, 171.

168. Ibid., 171.

169. Ibid., 243.

170. Ibid., 171.

171. Ibid., 243.

172. Anton M. Matytsin, *The Specter of Skepticism in the Age of Enlightenment* (Baltimore, MD: Johns Hopkins University Press, 2016), 2.

173. Here again, there are important connections between Hume and Haywood. See Garfield, *Concealed Influence*.

174. Bacon, *Novum Organum*, 76.

175. For the relationship between wonder and Enlightenment-era philosophy and fiction, see Kareem, *Eighteenth-Century Fiction and the Reinvention of Wonder*; Girten, "Trivial Wonders."

Chapter Four

Portions of chapter 4 appeared in Kristin M. Girten, "Charlotte Smith's Tactile Poetics," *The Eighteenth Century: Theory and Interpretation* 54, no. 2 (summer 2013): 215–30 and are reprinted with permission of the University of Pennsylvania Press.

1. Charlotte Smith, *Beachy Head*, in *The Poems of Charlotte Smith*, ed. Stuart Curran (Oxford: Oxford University Press, 1993), lines 687–88, page 246.

2. Alex Wetmore, "Sentimentalism and Sensibility," in *Mary Wollstonecraft in Context*, ed. Nancy E. Johnson and Paul Keen (Cambridge: Cambridge University Press, 2020), 258.

3. G. J. Barker-Benfield, *The Culture of Sensibility: Sex and Society in Eighteenth-Century Britain* (Chicago: University of Chicago Press, 1996), 296.

4. "Sensibility," *Oxford English Dictionary*.

5. "Sentimental," *Oxford English Dictionary*.

6. See for instance Elizabeth A. Dolan, "British Romantic Melancholia: Charlotte Smith's *Elegiac Sonnets*, Medical Discourse and the Problem of Sensibility," *Journal of European Studies* 33, no. 3–4 (December 12, 2003): 237–53; Claire Knowles, *Sensibility and Female Poetic Tradition, 1780–1860: The Legacy of Charlotte Smith* (Farnham, England: Ashgate Publishing, 2009); Erinç Özdemir, "Charlotte Smith's Poetry as Sentimental Discourse," *Studies in Romanticism* 50, no. 3 (Fall 2011): 437–73; Adrianne Wadewitz, "The Politically Engaged Child: Charlotte Smith's Children's Literature and the Discourse of Sensibility," in *Beyond Sense and Sensibility: Moral Formation and the Literary Imagination from Johnson to Wordsworth*, ed. Peggy Thompson (Lewisburg, PA: Bucknell University Press, 2015), 91–106.

7. Smith may thus be seen to join twentieth- and twenty-first-century theorists who have similarly emphasized the value of a feminist ethic of care. See for example, Virginia Held, *The Ethics of Care: Personal Political, and Global* (Oxford: Oxford University Press, 2006); Joan C. Tronto, *Moral Boundaries: A Political Argument for an Ethic of Care* (New York: Routledge, 1993).

8. Kevis Goodman, "Conjectures on Beachy Head: Charlotte Smith's Geological Poetics and the Ground of the Present," *English Literary History* 81, no. 3 (Fall 2014): 994.

9. On the botanical elements of Smith's poetry, see especially Theresa M. Kelley, "Romantic Exemplarity: Botany and 'Material' Culture," *Romantic Science: The Literature Forms of Natural History*, ed. Noah Heringman (Albany: State University of New York Press, 2003), 223–54; Judith Pascoe, "Female Botanists and the Poetry of Charlotte Smith," in *Revisioning Romanticism: British Women Writers, 1776–1837*, ed. Carol Shiner Wilson and Joel Haefner (Philadelphia: University of Pennsylvania Press, 1994), 193–209; Donelle R. Ruwe, "Charlotte Smith's Sublime: Feminine Poetics, Botany, and *Beachy Head*," *Prism(s): Essays in Romanticism* 7 (1999): 117–32.

10. Charlotte Smith, *The Collected Letters of Charlotte Smith*, ed. Judith Phillips Stanton (Bloomington: Indiana University Press, 2003), 283–84. Unfortunately, the collection this letter proposes was never produced.

11. Ibid., 283–84.

12. See Ann Bermingham, *Learning to Draw: Studies in the Cultural History of a Polite and Useful Art* (New Haven, CT: Yale University Press, 2000); Sam George, *Botany, Sexuality, and Women's Writing, 1760–1830: From Modest Shoot to Forward Plant* (Manchester: Manchester University Press, 2007); Ann B. Shteir, *Cultivating Women, Cultivating Science: Flora's Daughters and Botany in England, 1760 to 1860* (Baltimore, MD: Johns Hopkins University Press, 1996); Elise L. Smith, "Centering the Home-Garden: The Arbor, Wall, and Gate in Moral Tales for Children," *Children's Literature* 36 (2008): 24–48.

13. Charlotte Smith, *Rural Walks*, in *The Works of Charlotte Smith*, vol. 12, ed. Elizabeth A. Dolan (London: Pickering & Chatto, 2007), 47.

14. Smith's other educational works are: *Rambles Farther* (London, 1796), *Minor Morals* (London, 1798), *Letters of a Solitary Wanderer* (London, 1800), *Conversations Introducing Poetry* (London, 1804), and *A Natural History of Birds* (London, 1807).

15. Smith, *Rural Walks*, 28–29; 43–48; 86–89; 124; 127–29; 137.

16. Ibid., iii–iv.

17. Ibid., 31.

18. Ibid., 50–52.

19. Ibid., iii.

20. According to Loraine Fletcher, "[h]er writing for children soon began to pay well." Loraine Fletcher, *Charlotte Smith: A Critical Biography* (London: Palgrave, 1998), 228. On how Smith's financial pressures motivated her writing, see Fletcher, "Writing to Live," in *Charlotte Smith*, 78–128.

21. Smith, *Collected Letters*, 368.

22. Ibid., 368n2.

23. As Loraine Fletcher suggests, she sought out Michael Drayton's topographical epic poem *PolyOlbion* (1612) for inspiration as she was working on *Beachy Head*. Fletcher, *Charlotte Smith*, 327. Other important precursors and likely influences are Virgil's *Georgics*, John Denham's *Cooper's Hill* (1642), John Milton's *Paradise Lost* (1667), Alexander Pope's *Windsor-Forest* (1713), James Thomson's *The Seasons* (1726–30), Thomas Gray's "Elegy Written in a Country Churchyard" (1751), William Cowper's *The Task* (1785), and Erasmus Darwin's *The Loves of the Plants* (1789) and *The Temple of Nature* (1803). For a helpful overview of *Beachy Head*'s precursors, see Juan Christian Pellicer, "Virgil's Eclogues and Georgics in Charlotte Smith's *Beachy Head*," in *Romans and Romantics*, eds. Timothy Saunders, Charles Martindale, Ralph Pite, and Mathidle Skoie (Oxford: Oxford University Press, 2012), 161–82.

24. Smith, *Beachy Head*, line 23, page 217.

25. Ibid., lines 5–10, page 217.

26. Anne D. Wallace, "Picturesque Fossils, Sublime Geology? The Crisis of Authority in Charlotte Smith's *Beachy Head*," *European Romantic Review* 13, no. 1 (2002): 78.

27. Smith, *Beachy Head*, line 463, page 237.

28. Ibid., line 467, page 237.

29. Ibid., line 226, page 236.

30. James Sowerby, *English Botany* (London, 1790), 64–65, 22.

31. Smith, *Beachy Head*, line 461, page 237.

32. Ibid., line 25, page 218.

33. Ibid., line 514, page 239.

34. Ibid., lines 390–94, page 233.

35. As Goodman explains, "[f]or Smith and her contemporaries, the term 'conjecture' would have evoked the intellectual practice called 'conjectural history,' which crossed the several human and natural sciences of the Scottish Enlightenment." Goodman, "Conjectures," 995.

36. In the context of conjectural history, such overestimation manifests as the "present[ation of] a human construction as a natural phenomenon without marking or acknowledging its constructedness." Goodman, "Conjectures," 996.

37. Smith, *Beachy Head*, lines 419–22, page 235.

38. Ibid., line 524, page 239.

39. Bacon, *Novum Organum*, 8.149.

40. Ibid., 8.148.

41. Anton M. Matytsin, *The Specter of Skepticism in the Age of Enlightenment* (Baltimore, MD: Johns Hopkins University Press, 2016).

42. Richard Pulteney, *A General View of the Writings of Linnæus* (London, 1781), iv.

43. My emphasis; John Locke, *An Essay Concerning Human Understanding*, ed. Roger Woolhouse (London: Penguin Books), 584.

44. Locke, *Essay*, 584–85.

45. Francis Bacon, *The New Organon*, trans. and ed. Lisa Jardine and Michael Silverthorne (New York: Cambridge University Press, 2000), 222.

46. Thomas Sprat, *History of the Royal Society* [1667], ed. Jackson I. Cope and Harold Whitmore (St. Louis: Washington University Press, 1958), 66–73.

47. Steven Shapin, *A Social History of Truth: Civility and Science in Seventeenth-Century England* (Chicago: University of Chicago Press, 1994), xxvi.

48. Donna J. Haraway, *Modest_Witness@Second_Millenium.FemaleMan_Meets_OncoMouse: Feminism and Technoscience* (New York: Routledge, 1997), 27.

49. Pamela Clemit and Brad Scott, "Botanical Networking: Four Holograph Letters from Charlotte Smith to James Edward Smith," *Romanticism* 26, no. 1 (2020): 1–12. According to Clemit and Scott, though "Charlotte Smith did not find employment as a botanical colourist, her reciprocal interaction with J. E. Smith yielded other benefits. The correspondence appears to have been a catalyst for her later botanical writings, which show an increasingly intellectual and scientific engagement with nature." Clemit and Scott, "Botanical Networking," 3.

50. Smith, *Beachy Head*, line 446, page 236.

51. Ibid., line 390–92, page 233.

52. Ibid., lines 1–4, page 217.

53. On the history of the eighteenth-century British prospect poem, see especially

John Barrell, *English Literature in History: An Equal Wide Survey* (London: Hutchinson, 1983); Helen Deutsch, "Elegies in Country Churchyards: The Prospect Poem in and around the Eighteenth Century," in *The Oxford Handbook of the Elegy*, ed. Karen Weisman (Oxford: Oxford University Press, 2010), 187–205.

54. In contrast, "beautiful objects are comparatively small." Moreover, "the ideas of the sublime and the beautiful stand on foundations so different, that it is hard, I had almost said impossible, to think of reconciling them in the same subject." Edmund Burke, *A Philosophical Enquiry into the Origin of Our Ideas of the Sublime and Beautiful* [1757], ed. James T. Boulton (Notre Dame, IN: Notre Dame Press, 1968), 72, 113–14.

55. Immanuel Kant, *The Critique of Judgment* [1790], ed. and trans. James Creed Meredith (Oxford: Oxford University Press, 1952), 97.

56. Some particularly influential and illustrative embodiments of this association are John Denham, "Cooper's Hill," in *Poems and Translations* (London, 1668); Richard Blackmore, *Creation* (London, 1712); James Thomson, *The Seasons* (London, 1730); John Dyer, *The Fleece* (London, 1757); William Cowper, *The Task* (London, 1785).

57. My emphasis; Burke, *Enquiry*, 40.

58. What scholars regularly see as a contradiction in Burke's writings between his celebration of the sublime and his abhorrence of the French Revolution (which was, by its advocates, regularly portrayed as sublime) is attributable to his insistence that the sublime is contingent on distance. For Burke, the French Revolution was only horrifying and *not* sublime precisely because it pressed too close. Frances Ferguson, *Solitude and the Sublime: The Romantic Aesthetics of Individuation* (New York: Routledge, 1992); Tom Furniss, *Edmund Burke's Aesthetic Ideology: Language, Gender, and Political Economy in Revolution* (Cambridge: Cambridge University Press, 1993).

59. Kant, *Critique of Judgment*, 112.

60. Smith, *Beachy Head*, lines 368–75, 232.

61. "Keen," *Oxford English Dictionary*.

62. Goodman, "Conjectures," 998.

63. Pascoe, "Female Botanists."

64. Burke, *Enquiry*, 72–73.

65. Smith, *Beachy Head*, lines 12–17, page 217.

66. *Ibid.*, lines 346–63, page 231.

67. Ibid., line 591, page 242.

68. Smith was certainly distinctive as a botanist-poet but was far from alone in composing poetry informed by botanical practice and knowledge. Erasmus Darwin set an important foundation for such a technique, and Smith was one among several poets to follow his example. On literary adaptations of botany, see especially Shteir, *Cultivating Women*; George, *Botany, Sexuality, and Women's Writing*; Sam George, "Carl Linnaeus, Erasmus Darwin, and Anna Seward: Botanical Poetry and Female Education," *Science and Education* 23, no. 3 (2014): 673–94; Theresa Kelley, *Clandestine Marriage: Botany and Romantic Culture* (Baltimore, MD: Johns Hopkins University Press, 2012); Amy M. King, *Bloom: The Botanical Vernacular in the English Novel* (Oxford: Oxford University Press, 2003); Patricia Fara, *Erasmus Darwin: Sex, Science, and*

Serendipity (Oxford: Oxford University Press, 2012). For an extended study of Smith's botanical literary adaptations, see Jacqueline Labbe, *Charlotte Smith: Romanticism, Poetry and Culture of Gender* (Manchester: Manchester University Press, 2003).

69. Smith, *Beachy Head*, lines 479–92, pages 237–38.

70. Ibid., line 481, page 238.

71. Robert Boyle, "Proëmical Essay . . . with some Considerations touching Experimental Essays in General," in *The Works of the Honourable Robert Boyle*, vol. 1, 2nd ed., ed. Thomas Birch (London: J. & F. Rivington, 1772), line 318.

72. Various scholars have explored connections between the poetry of Charlotte Smith and Darwin. See especially Melissa Bailes, "Linnaeus's Botanical Clocks: Chronobiological Mechanisms in the Scientific Poetry of Erasmus Darwin, Charlotte Smith, and Felicia Hemans, *Studies in Romanticism* 56, no. 2 (Summer 2017): 223–52; Rosalind Powell, "Linnaeus, Analogy, and Taxonomy: Botanical Naming and Categorization in Erasmus Darwin and Charlotte Smith," *Philological Quarterly* 95, no. 1 (2016): 101–24.

73. Tita Chico, *The Experimental Imagination: Literary Knowledge and Science in the British Enlightenment* (Stanford, CA: Stanford University Press, 2018), 25.

74. Abraham Cowley, "To the Royal Society," in *The History of the Royal-Society of London for the Improving of Natural Knowledge*, by Thomas Sprat (London, 1667).

75. *The Record of the Royal Society of London for the Promotion of Natural Knowledge*, 4th ed. (London: Morrison & Gibb, 1940), 290.

76. Erasmus Darwin, *The Loves of the Plants* (London, 1789), lines 39–44, page 3.

77. Ibid., 3.

78. Smith, *Beachy Head*, lines 363–67, pages 231–32.

79. This is only one of several ways in which Smith's *Beachy Head* deviates from and implicitly challenges Darwin's poetics. On their differences, see especially Goodman's analysis, which is informed by Alan Bewell's study of Darwin's "Cosmopolitan Nature." Goodman, "Conjectures," 993; Alan Bewell, "Erasmus Darwin's Cosmopolitan Nature," *English Literary History* 76, no. 1 (Spring 2009): 19–48. See also Dahlia Porter's discussion of Smith's *Conversations Introducing Poetry, Chiefly on Subjects of Natural History*. Porter argues that "Smith . . . presents her collection of poems as a counterpoint to books like Darwin's *Loves [of the Plants]* that cover over the violent histories of colonization with animated figures and personified plants." Dahlia Porter, "Specimen Poetics: Botany, Reanimation, and the Romantic Collection," *Representations* 139 (Summer 2017): 81.

80. Smith, *Beachy Head*, line453, page 236.

81. Henry Baker, *The Universe. A Poem.* (London, 1734), 7–8.

82. Smith, *Beachy Head*, lines 4–17, page 217.

83. Baker, *The Universe*, 1.

84. Henry Baker, "An Account of some uncommon fossil Bodies," Read March 29, 1753, *Philosophical transactions: giving some accompt of the present undertakings, studies and labours of the ingenious in many considerable parts of the world*, vol. 48, ed. Henry Oldenburg (London, 1754), 118.

85. Baker's papers in the *Philosophical Transactions* thus challenge Al Coppola's claim that, by the third decade of the eighteenth century, "the virtuosa came to embody the ideal witness for producing scientific truth in the first place," thereby altogether supplanting the modest witness. Al Coppola, *The Theater of Experiment: Staging Natural Philosophy in Eighteenth-Century Britain* (Oxford: Oxford University Press, 2016), 164. While immodest practices of witnessing may have, as Coppola shows, distinguished a range of popular philosophical works, detachment and neutrality—hallmarks of Baconian modesty—persisted within those serious philosophical works published by male virtuosi for male virtuosi.

86. Chico, *Experimental Imagination*, 31.

87. Henry Power, *Experimental Philosophy in Three Books*, vol. 1 (London, 1664), 1:6–7.

88. Sprat, *The History of the Royal Society*, 335.

89. Chico, *Experimental Imagination*, 41; Sprat, *The History of the Royal Society*, 335. See also Lorraine Daston, "Enlightenment Fears, Fears of Enlightenment," in *What's Left of Enlightenment? A Postmodern Question*, ed. Keith Michael Baker and Peter Hanns Reill (Stanford, CA: Stanford University Press, 2001), 118, 120.

90. Smith, *Beachy Head*, lines 85–86, page 220.

91. As Goodman observes, the "sense of place" that Smith portrays in *Beachy Head* "differs from the ordered, stable landscape and 'equal wide survey' of the prospect poem described by John Barrell." Goodman, "Conjectures," 992; Barrell, *English Literature in History*. Smith, *Beachy Head*, lines 376–89, pages 232–33.

92. Analyzing it in the context of "conjectural history," Kevis Goodman portrays the shared materiality that distinguishes the poem's setting as a "*conjectus* of physical nature" and, thus, a "composite quality." Goodman, "Conjectures," 996.

93. On the Romantic "material sublime," see especially John G. Pipkin, "The Material Sublime of Women Romantic Poets," *SEL* 38 (1998): 597–619; Steve Vine, "Blake's Material Sublime," *Studies in Romanticism* 41 (Summer 2002): 237–57.

94. Smith, *Beachy Head*, lines 376–89, pages 232–33.

95. Ibid., line 382, page 232.

96. I would suggest that further consideration of Smith's conjectural techniques may reveal new connections between her poetics and those of Wordsworth and Keats who also, as Geoffrey Hartman has recognized in an influential analysis of "The Solitary Reaper," employed surmise as a poetic technique. Geoffrey Hartman, "'The Solitary Reaper,'" in *Wordsworth's Poetry, 1787–1814* (New Haven, CT: Yale University Press, 1971), 3–18.

97. On the dichotomous gendering of beauty and the sublime, see especially Ian Balfour, "Torso: (The) Sublime Sex, Beautiful Bodies, and the Matter of the Text," *Eighteenth-Century Studies* 39, no. 3 (2006), 323–36. As Balfour acknowledges, "the distinction [Burke's] *Enquiry* thematizes between the male sublime and the female beautiful . . . has extraordinary staying power in the period and is rarely challenged on a theoretical level." In fact, Balfour goes on to assert, "It is, as far as I can tell, almost only in women's fiction of the period, as in the Gothic romances of Radcliffe,

Dacre and Mary Shelley that the paradigm I have outlined undergoes a critique or a troubling, with female heroines possessing an articulate sense of the sublime, presented in texts written by women and thus doubly beyond the Burkean/Kantian paradigm." However, I would suggest that more work needs to be done to appreciate the challenges eighteenth-century and Romantic female poets posed to Burke and Kant. Balfour neglects to include Charlotte Smith in his study, but as the present chapter demonstrates, her poetry offers a strong counterpoint to Burke and Kant. Balfour, 332. On women's fiction and the eighteenth-century sublime, see also Barbara Claire Freeman, *The Feminine Sublime: Gender and Excess in Women's Fiction* (Berkeley, CA: University of California Press, 1997).

98. Dionysius Longinus, *On the Sublime*, ed. and trans. William Smith (Dublin, 1740), 21.

99. Ibid., 21.

100. Balfour, "Torso," 325.

101. Burke, *Enquiry*, 59.

102. Ibid., 59. John Milton, *Paradise Lost*, Book 2 (London, 1667), 666–73 (misquoted).

103. John Milton, *Paradise Lost*, Book 2, 648–59.

104. Burke, *Enquiry*, 60; Kant, *Critique of Judgement*, 104–105.

105. Smith, *Beachy Head*, line 686, 246.

106. Ibid., lines 687–88, page 246.

107. Smith, Advertisement to *Beachy Head*, page 215.

108. Smith, *Beachy Head*, line 675, page 245.

109. Ibid., lines 693–97, page 246.

110. Ibid., line 686, 246; lines 672–74, page 245.

111. Ibid., lines 717–26, page 247.

112. My emphasis; ibid., lines 726–31, page 247.

113. Barker-Benfield, *Culture of Sensibility*, 262.

114. Dolan argues that "in her negotiation of the gendered medical and cultural definitions of melancholia in her poetry, Smith invented a model for managing emotion that had an enormous influence on what we now call Romantic poetry." Dolan, "British Romantic Melancholia," 239–40.

115. Smith, Preface to *Elegiac Sonnets*, *Poems*, 3.

116. My emphasis; Smith, "Sonnet XXI. Supposed to be written by Werter," *Poems*, 26–27.

117. Eric Parisot, "Suicide Notes and Popular Sensibility in the Eighteenth-Century British Press." *Eighteenth-Century Studies* 47, no. 3 (Spring 2014): 277–91; Inger Sigrun Brodey, "Masculinity, Sensibility, and the 'Man of Feeling': The Gendered Ethics of Goethe's Werther," *Papers on Language and Literature: A Journal for Scholars and Critics of Language and Literature* 35, no. 2 (Spring 1999): 115–40.

118. David Hume, *A Treatise of Human Nature*, ed. L. A. Selby-Bigge; 2nd ed., ed. P. H. Nidditch (Oxford: Oxford University Press, 1978), 634. This correspondence is particularly striking if "determination" is defined as "the determining of bounds or

fixing of limits; delimitation; definition; a fixing of the extent, position, or identity (*of* anything)." "Determination," *Oxford English Dictionary Online.*

119. I have chosen to assign the speaker of this sonnet a masculine gender because Smith includes this sonnet in her novel *Emmeline* and attributes it to her male character Godolphin. Charlotte Smith, *Emmeline*, vol. 3 (London, 1788), 256.

120. For reviews that emphasize Smith's allegiance with feeling, see for example Anonymous, "Art. 21. Elegiac Sonnets, and Other Essays," *English Review, or, An Abstract of English and Foreign Literature, 1783–1795* 4 (September 1784): 231; Tobias George Smollett, ed., "Elegiac Sonnets," *The Critical Review, or, Annals of Literature* 61 (June 1786): 467–68; J. T., "An Essay on the English Sonnet; Illustrated by a Comparison between the Sonnets of Milton and Those of Charlotte Smith," *Universal Magazine of Knowledge and Pleasure, June 1747–Dec. 1803* 91 (December 1792): 408–14.

121. Smith, *Beachy Head*, lines 50–60, pages 218–19.

122. Wetmore, "Sentimentalism and Sensibility," 262. For other scholarship that complicates this traditional view, see also Simon Swift, "Mary Wollstonecraft and the 'Reserve of Reason,'" *Studies in Romanticism* 45 (Spring 2006): 3–24; Barbara Taylor, *Mary Wollstonecraft and the Feminist Imagination* (Cambridge: Cambridge University Press, 2003); John Whale, *Imagination under Pressure, 1789–1832: Aesthetics, Politics, and Utility* (Cambridge: Cambridge University Press, 2000).

123. My emphasis on "emotion that reason deepens"; Mary Wollstonecraft, *A Vindication of the Rights of Men* (Oxford: Oxford University Press, 1994), 54.

124. Smith, "Sonnet XXI. Supposed to be written by Werter," *Poems*, 26.

125. Smith, *Beachy Head*, lines 690–91, page 246.

126. Ibid., lines 700–709, pages 246–47.

127. Ibid., lines 710–16, page 247.

128. As Goodman explains, "The mixed and aggregated layers making up the rock formation called Beachy Head constitute a giant *conjectus* of physical nature, and it is this composite quality of the earth that throughout the poem is occasion for delight, not censure. Similarly—and just as importantly—the poetic formation also called *Beachy Head* is an unabashedly heterogeneous assembly, mixing lyric song and narrative verse, myths and histories of Britain with details from botany and other sciences." Goodman, "Conjectures," 996. See also Jacqueline Labbe, "The Hybrid Poems of Smith and Wordsworth: Questions and Disputes," *European Romantic Review* 20 (2009): 219–26; Dahlia Porter, "From Nosegay to Specimen Cabinet: Charlotte Smith and the Labour of Collecting," *Charlotte Smith in British Romanticism*, ed. by Jacqueline Labbe (London: Pickering & Chatto, 2008), 29–44.

129. As Adela Pinch argues, though "the romantic period and the era that precedes it often appear to us as eras in which people increasingly learned to claim their emotions as the guarantors of their individuality," certain works of the long Enlightenment nevertheless "reveal the period's concomitant tendency to characterize feelings as transpersonal, as autonomous entities that do not always belong to individuals but rather wander extravagantly from one person to another." Adela Pinch, *Strange Fits of*

Passion: Epistemologies of Emotion, Hume to Austen (Stanford, CA: Stanford University Press, 1996), 3.

130. The ontology of kinship I identify in Smith's verse is suggestive of the relationship between human and nature—in which nature appears as "a *female friend*, a sister"—that Anne Mellor sees in Smith's writing. However, where Mellor portrays this relationship as anthropocentric, I argue for its ecological inclusivity. Anne K. Mellor, *Romanticism and Gender* (New York: Routledge, 1993), 97.

131. Smith, *Minor Morals*, 2nd ed., *Works*, vol. 12, 219–20.

132. Ibid., 269–70.

133. Ibid., 270.

134. Smith, *Beachy Head*, lines 461–69, page 237.

135. The terms "ravage" and "mutilate" are taken from the dramatically opposed sort of tactile encounter with nature that William Wordsworth portrays in his poem "Nutting" (1800). William Wordsworth, *The Poetical Works of William Wordsworth*, ed. E. De Selincourt (Oxford: Oxford University Press, 1944), 2: lines 45, 50, pages 211–12.

136. Fletcher, *Charlotte Smith*, 62–63.

137. Fletcher, *Charlotte Smith*, 338.

138. "Charlotte Smith [*née* Turner]," *Oxford Dictionary of National Biography.*

139. Qtd. in Fletcher, *Charlotte Smith*, 38.

Coda

1. Stacy Alaimo and Susan J. Hekman, *Material Feminisms* (Bloomington: Indiana University Press, 2008), 6.

2. Ibid., 3–4.

3. Ibid., 2–3.

4. Jane Bennett, *Vibrant Matter: A Political Ecology of Things* (Durham, NC: Duke University Press, 2010), 113.

5. Ibid., 119.

6. Ibid., 113.

7. Ibid., 116.

8. Ibid., 116.

9. Ibid., 113.

10. Charles Taylor, *A Secular Age* (Cambridge: Cambridge University Press, 2007), 37–38.

11. Bennett, *Vibrant Matter*, 65, 4–17.

12. Ibid., 117.

13. See for instance Virginia Held, *The Ethics of Care: Personal Political, and Global* (Oxford: Oxford University Press, 2006); Joan C. Tronto, *Moral Boundaries: A Political Argument for an Ethic of Care* (New York: Routledge, 1993).

14. Mary Shelley, introduction to *Frankenstein*, 3rd ed. (1831), *Frankenstein*, ed. J. Paul Hunter, 2nd ed. (New York: W. W. Norton & Company, 2012), 169.

15. Shelley, *Frankenstein* (1818), ed. J. Paul Hunter, 71, 74, 97.

16. See for instance Helena Feder, "'A Blot upon the Earth': Nature's 'Negative' and

the Production of Monstrosity in *Frankenstein,*" *Journal of Ecocriticism* 2, no. 1 (Jan 2010): 55–66; Timothy Morton, "*Frankenstein* and Ecocriticism," in *The Cambridge Companion to* Frankenstein, ed. Andrew Smith (Cambridge: Cambridge University Press, 2016), 143–57; Shalon Noble, "An Uncertain Spirit of an Unstable Place: *Frankenstein* in the Anthropocene," in *Romantic Ecocriticism: Origins and Legacies*, ed. Dewey W. Hall (Lanham, MD: Lexington Books, 2016), 123–38; Paul Outka, "Posthuman/Postnatural: Ecocriticism and the Sublime in Mary Shelley's *Frankenstein*," in *Environmental Criticism for the Twenty-First Century*, ed. Stephanie LeMenager, Teresa Shewry, and Ken Hiltner (New York: Routledge, 2011), 31–48.

17. Shelley, *Frankenstein*, 157.

18. Ibid., 157.

19. Ibid., 159.

20. My emphasis; ibid., 159.

21. Ibid., 159.

22. Ibid., 159.

23. Ibid., 161.

24. Ibid., 161.

25. Mary Wollstonecraft, *A Vindication of the Rights of Men* (Oxford: Oxford University Press, 1994), 54.

26. Ibid., 54; Shelley, *Frankenstein*, 159.

27. Wollstonecraft, *Vindication*, 54; Shelley, *Frankenstein*, 159.

28. On the "separation of the spheres" in the era of Enlightenment, see especially Amanda Vickery, "Golden Age to Separate Spheres? A Review of the Categories and Chronology of English Women's History," *The History Journal* 36, no. 2 (1993); Karen O'Brien, *Women and Enlightenment in Eighteenth-Century Britain* (Cambridge: Cambridge University Press, 2009); Robert B. Shoemaker, *Gender in English Society, 1650–1850: The Emergence of Separate Spheres?* (London: Longman, 1998); Harriet Guest, *Small Change: Women, Learning, Patriotism, 1750–1810* (Chicago: University of Chicago Press, 2000).

29. Donna J. Haraway, *Modest_Witness@Second_Millenium.FemaleMan_Meets_OncoMouse: Feminism and Technoscience* (New York: Routledge, 1997), 24.

30. Lesley Head and Theresa Harada, "Keeping the Heart a Long Way from the Brain: The Emotional Labour of Climate Scientists," *Emotion, Space and Society* 24 (2017): 40.

31. Ibid., 37. As Head and Harada ("Keeping the Heart") acknowledge,

> It is important to emphasize here that the sample is already skewed towards personal resilience and the ability to survive and succeed in a competitive academic environment of limited resources. Anyone working in a university science context would recognize this male-dominated sample. The number of opportunities shrink at each career stage, creating a funneling effect, with even the early career participants here already survivors of such a system. This perhaps reinforces how emotional management strategies have contributed to attaining their current positions. (39)

32. Ibid., 39.

33. Ibid., 40.

34. Robin Wall Kimmerer, *Gathering Moss: A Natural and Cultural History of Mosses* (Corvallis: Oregon State University Press, 2003), vii.

35. Ibid., vii.

36. Cary Wolfe, *What Is Posthumanism?* (Minneapolis: University of Minnesota Press, 2010), xv.

37. Ibid., xv.

38. Ibid., xv.

INDEX

Individual works are listed under their author's name.

Kolibri 360 GmbH
Ferdinand-Jühlke-Straße 7
99095 Erfurt, Deutschland
produktsicherheit@kolibri360.de